Lean Design Management

The fields of design management and lean construction appear to be developing independently. This volume brings together authors from four continents to argue that lean thinking should be integral to design management in the fields of architecture, engineering and construction (AEC).

The publication brings together a variety of perspectives on lean design management as experienced in Africa, South America, Australasia and Europe. Themes covered include:

- Lean thinking and flexible building solutions
- Quality and flow of information in 14 sub-projects of a major airport project
- Ghanaian consultants' perspective on process waste
- Target costing and its application to social housing projects in Brazil
- Concept of 'first' and 'last' value, drawing on social housing projects in Chile
- Development of a lean design management model specifically for remote sites

The chapters offer a mix of theoretical materials and empirical research findings, providing a unique insight into aspects of lean design management.

This book was published as a special issue of *Architectural Engineering and Design Management*.

Stephen Emmitt is Professor of Architectural Technology in the School of Civil and Building Engineering at Loughborough University, UK. He is an architect and has written widely on design management in AEC.

Lean Design Management

Edited by
Stephen Emmitt

Routledge
Taylor & Francis Group
LONDON AND NEW YORK

First published 2012
by Routledge
2 Park Square, Milton Park, Abingdon, Oxfordshire OX14 4RN

Simultaneously published in the USA and Canada
by Routledge
711 Third Avenue, New York, NY 10017

First issued in paperback 2014

Routledge is an imprint of the Taylor & Francis Group, an informa business

This book is a reproduction of *Architectural Engineering and Design Management*, vol.7, issue 2. The Publisher requests to those authors who may be citing this book to state, also, the bibliographical details of the special issue on which the book was based.

British Library Cataloguing in Publication Data
A catalogue record for this book is available from the British Library

ISBN13: 978-1-84971-431-0 (hbk)
ISBN13: 978-1-138-85337-9 (pbk)

Typeset in Helvetica
by Taylor & Francis Books

Publisher's Note

The publisher would like to make readers aware that the chapters in this book may be referred to as articles as they are identical to the articles published in the special issue. The publisher accepts responsibility for any inconsistencies that may have arisen in the course of preparing this volume for print.

Contents

Introduction

Stephen Emmitt

Professor of Architectural Technology, Loughborough University, UK

The driver behind this special edition of *Architectural Engineering & Design Management* is curiosity. Curiosity as to why lean design management has not been tackled with the same enthusiasm as lean construction and curiosity as to what lean design management can contribute to our collective understanding of design management in architecture–engineering–construction (AEC).

As a practising architect and design manager I was first introduced to the lean thinking concept in the early 1990s by one of our clients, who rather unceremoniously dumped a copy of Womack *et al.*'s book *The Machine that Changed the World: The Story of Lean Production* on my desk and pronounced 'this is what I want'. At that time we were working with many initiatives to improve the quality of the service and the buildings delivered to our clients, but not with lean ideas. Although the contents of the book required a lot of interpretation to make it comprehensible and relevant to an architectural practice working in a transient and dynamic project environment, the core message got through and we started to implement some aspects of lean production into our office and project environment. However, it was not until 2003, when I moved to Scandinavia as the Hoffmann Professor of Innovation and Management in Building at the Technical University of Denmark (DTU), that I started to look in detail at lean construction, with a particular focus on the value of design and the importance of lean design management. Working with consultants and contractors in Denmark allowed me the opportunity to see the development and practical application of lean thinking: in particular, the gradual shift of focus from the construction site to the earlier project stages, and then to the implementation of a lean culture in the organizations contributing to the projects. More recently, on my return to the UK, I have been involved in a knowledge transfer project that has been highly successful in bringing about a lean culture within a design and build contracting organization. Again, it has been interesting to experience the shift in focus upstream, starting with site activities and moving towards lean design management to better address the value and values aspects of multidisciplinary projects.

Working with doctoral students at DTU and Loughborough University has helped to highlight the paucity of theoretical work in design management and specifically lean design management. There also appears to be a lack of empirical evidence on the application of lean design management in practice. Collectively this experience has convinced me that we need to devote more attention to the creation of lean organizations and the management of the early stages of lean projects. By doing so it is possible to mitigate many of the familiar challenges we still encounter on a daily basis during the construction phase and to add value to all project stakeholders, be they the funder, user or larger society in which the completed artefact sits. The point of departure for this special edition is that lean thinking should be integral to design management in the AEC sectors. Building on previous special editions *Aspects of Design Management* (Volume 3:1, 2007) and *Design Management for Sustainability* (Volume 5:1&2, 2009), this publication brings together a variety of perspectives on lean design management from authors located in four continents: Africa, South America, Australasia and Europe. The articles provide a unique insight into some of the elements of lean design management and, intriguingly, open the door for further research and application.

In the first article, Hansen and Olsson take an integral approach to the subject exploring the concept of layered projects and layered processes. In so doing, they make an important link between lean thinking and flexible building solutions, drawing on large hospital projects in Norway and Switzerland to make their point. In this article, the concept of lean design management is fully integrated with the use of the building and hence customer value. The authors confront some of the contradictory characteristics of projects, concluding that there is a need for flexibility and a need to see the design process and the design as integrated elements.

Quality and flow of information has long been a cause for concern in AEC projects and in many respects lies at the heart of all projects, be they lean or otherwise. Information flows in multidisciplinary civil engineering design teams using lean measures is the theme of the second article. Tribelsky and Sacks draw on a large data set to analyse 14 sub-projects of a major airport project. They found a positive correlation between the quality of information flow and the quality of design documentation. Unstable information flows were associated with unpredictable project outcomes, helping to emphasize the importance of flow – an underlying lean principle.

The geographical spread of lean is nicely illustrated in the third article, which reports a questionnaire survey of Ghanaian consultants and their perspective on process waste. Kpamma and Adjei-Kumi found a relatively low level of recognition of the various sources of waste, and a low awareness of waste reduction tools and techniques. The authors conclude that much is still to be done to raise awareness of waste and lean thinking in Ghana and point to initiatives of how this can be achieved. These observations appear to be in line with other countries, where lean has been, or is, slowly but surely permeating into everyday practice.

Interest in target costing is growing quite rapidly in the AEC sector, and could form an important element of a lean design management strategy. An investigation into the adoption of target costing on Brazilian social housing projects by Jacomit and Granja provides an extensive introduction to target costing for readers unfamiliar with the concept. Their extensive and balanced overview highlights a number of opportunities and barriers for the application of target costing. They also posit more efficient and proactive ways to conceive and deliver social housing projects that deliver more value to stakeholders and users.

From target costing, the focus naturally moves to the concept of value. In the penultimate article, Pasquire and Salvatierra-Garrido introduce the concept of 'first' and 'last' value, once again drawing on social housing projects, but this time from Chile. Here the argument is for elevating social issues into design management and for the lean design manager to take a whole life systems view. Although this article reports work in process, the work helps to clarify notions of value and extends the argument into a social, one could argue wider, knowledge domain.

In the final article, Kestle *et al.* reflect on recently completed doctoral work, describing the development of a lean design management model specifically for remote sites. The authors demonstrate the integration of lean design and design management thinking before describing the research into remotely located collaborative international projects on environmentally sensitive world heritage sites. An exploratory design management model is described and tested, with the finding that four factors are key to remote sites: value generation, knowledge integration, process integration and timely decision-making.

As stated earlier, one of the reasons for developing this special edition was curiosity, and the six articles published within these covers do go some way to further our understanding of lean design management, and hence a wider understanding of design management in AEC. Despite the geographical spread of the authors, some common understanding and some common themes have emerged. In very broad terms, the common themes relate to process, people, product and performance. In terms of understanding, it appears that lean thinking could (one might argue should) be part of an effective design management strategy, regardless of project context. The challenge for educators, researchers and practitioners is to integrate lean thinking into everyday design management practice.

I would like to acknowledge the great amount of work undertaken by the reviewers for this special edition and their commitment to academic rigour. A very enthusiastic response to the initial call for articles provided more work than anticipated, although the blind peer review process reduced the number of publishable articles to six. The articles contained here are a result of the authors' dedication to their subject and their positive reaction to the reviewers' detailed and insightful comments. On behalf of the authors, I would like to extend my gratitude to the reviewers, the experts behind the scenes, who have helped to bring about a stimulating and thought-provoking special edition.

Layered Project–Layered Process: Lean Thinking and Flexible Solutions

Geir K. Hansen and Nils O. E. Olsson

Department of Architectural Design and Management, Faculty of Architecture and Fine Art, Norwegian University of Science and Technology, N-7491 Trondheim, Norway

Abstract

This article elaborates on lean thinking in design management. Lean philosophy focuses on minimizing waste and maximizing value. We discuss how lean concepts can be understood and how this can affect the design process. In particular, we examine how flexibility can be achieved through the use of a layered design process and a layered building design. The design process has two main outcomes: input to the production and/or construction process, and input to the decision process. These two outcomes have a number of contradictory characteristics. Traditionally, production processes call for early decisions and control of cost, time and risks. Typically, this has been the engineers' or the contractors' perspective. In contrast, the decision process typically strives to postpone final commitments and design freeze in order to keep options open. Lean thinking shows that the two approaches do not have to be contradictory. Based on case studies of two hospital projects, we discuss flexibility in the design process and design in relation to lean principles. The article demonstrates the importance of seeing the design process and design model as two integrated parts. Lean thinking can serve as an enabler by focusing on reducing waste and increasing the customer's value.

INTRODUCTION

For many years, the building industry has been accused of poor performance on traditional indicators such as time, cost and quality, and of not being able to meet clients' needs, requirements and expectations. Building laws and regulations in many countries have traditionally focused on health, environment and safety, but in recent years there has been more focus on reducing faults and risks in order to increase the quality of the finished product. The building industry itself has developed quality systems and routines focusing on 'zero faults' and 'do it right the first time', thereby creating a growing market for project managers (Winch, 2010).

Lean thinking has been regarded as a success strategy in mass production, realizing cost and time savings, while simultaneously improving competitive positions in the market. The idea of lean thinking originated in Japanese manufacturing concepts (particularly the Toyota Production System) and has spread to other areas such as software development, logistics, service industries and construction. Lean thinking has also been introduced into the building construction sector through lean construction (Tommelein, 1998; Koskela, 1992, 2000). Similar principles as those in lean construction are also found in modern approaches to the development of IT systems, which include functional specifications and iterative development. These have been described by, for example, Poppendieck and Poppendieck (2003) and Boehm and Turner (2003).

Reducing manufacturing costs is desirable to achieve a measurable increase in profits, which is also essential for the building industry. A project's ability to produce its immediate outcome can be measured in terms of efficiency. It is a matter of doing things right and producing project outputs in terms of the agreed scope, quality, cost and time. This can be described as an internal measure. The longer term effects of a project can be measured in terms of effectiveness – in other words, in doing the right things. This is an external measure. OECD (2000) and Eikeland (2001) relate effectiveness to how the results of a project contribute to added value for owners and users. Emmitt *et al.* (2004) state that many of the problems that occur in the construction phase of a project are the result of ineffective communication and decision making in the design phase.

Most of the costs and much of the quality in a construction project are locked in long before production launch, and therefore the design process will be crucial not only to 'do things right' but also, more importantly, to 'do the right things'. Lean philosophy, with its focus on minimizing waste and maximizing value, should therefore, according to Emmitt *et al.* (2004), be applied as early as possible in the building process.

There are several approaches to lean design and lean design management. One approach is to streamline the design process itself in order to improve efficiency. Another approach is to provide design solutions for more effective construction. A third approach, which concerns the client's perspective, is to ensure that the design process reflects the client's needs and expectations, thereby adding value to the users and hence to the client's business. In this context, the value relates to how the use of the building supports the core business. The design process has two main aims or outcomes: input to the production and input to the managerial decision process, using the design as a decision-making tool. According to Blyth and Worthington (2010), the two outcomes have a number of contradictory characteristics. The production process calls for early decisions and control of cost, time and risks, traditionally focusing on the result or product. Typically, this has been the engineers' or the contractors' perspective. In contrast, Eikeland (2001) describes how the decision process calls for the last responsible moment in order to keep options open, focusing on the effect of the product or building, and representing the user's or the client's perspective. Traditionally, the two perspectives have been seen as contradictory (Olsson, 2006). However, design management approaches such as set-based design (Ward *et al.*, 1995) and lean construction (Koskela, 1992) have shown that the two approaches do not need to be, nor should be, conflicting.

One of the main challenges in the design and construction process is that neither all of the required information may be available nor the necessary decisions made in order to finalize the design. Instead, the design is used to develop possible solutions giving new understanding and input to the further process. In addition, the premises are often changed throughout the process. Set-based design (Ward *et al.*, 1995; Sobek *et al.*, 1999; Nahm and Ishikawa, 2005) focuses on keeping the design space open for as long as possible. This enables a project to develop a set of possible solutions to product design and production design problems, and for decisions to be made at the last responsible moment. The project team, including suppliers, may develop a number of design options and then choose the most appropriate.

Further, it is important to distinguish clearly between the characteristics and goals of designing and those of construction or making. Designing is an iterative process, a cycle of discovery, where finding and evaluating multiple solutions to a problem is regarded as adding value. Construction or making is a more sequential and linear process, where iterations often create waste and do not add value. The early phase of the design process can be described as both a problem-seeking and problem-solving process (Peña, 1987). It is a process of finding and evaluating multiple possible solutions to a problem, where iterations are necessary to be innovative, in order to find adequate concepts and designs. The approach of 'do it right the first time', which is an important slogan in production, should be applied with caution in design, as design freeze may be postponed until the last responsible moment. This means that decisions must be made within the lead time for realizing alternatives. A decision should not be made until it has to be made, in order to allow for flexibility (Ballard and Howell, 2003). According to Ballard (2000), the last responsible moment is the 'point at which failing to make the decision eliminates an alternative'. Emmitt *et al.* (2004) argue for a management framework that allows and encourages creativity where it is most valuable, that is, at the conceptual stage when values are discussed and agreed. Mikkelsen and Riis (2003) underline the need for both a flexible process and flexible and robust designs enabling changes, adjustments and development throughout a project.

This article discusses flexibility from different perspectives, including how a layered design process related to the concept of a layered building product can give a more flexible and open process that will be beneficial for the development process in design, right-in-time design solutions for construction and latest responsible moment for the client for decision making. Lean is considered from both an efficiency and effectiveness perspective.

The article is based on literature and theory on lean thinking, flexibility in construction projects and flexibility of buildings, with examples drawn mainly from two large European complex building projects.

LEAN THINKING AND FLEXIBILITY

The uncertainty of a decision in a project is commonly described as 'the gap between the amount of information needed to perform a task and the amount of information already possessed by the organisation' (Galbraith, 1973, p.5). Lean design and lean processes are a response to both the need to reduce uncertainty and the need for efficiency.

Olsson (2008) has proposed a distinction between internal and external project flexibility: project internal flexibility relates to flexibility within defined scope – *how* requirements will be met; while project external flexibility relates to adjustments of scope – *what* requirements will be met. Lean has a similar dual perspective in a design and construction context. From an external perspective, lean relates to customer value. The functionality or usability with the highest customer value is given priority. From an internal perspective, lean focuses on executing the delivery process with as little waste as possible.

Project internal flexibility has its roots in an efficiency perspective. According to Cremieux *et al.* (2005), flexibility may be characterized as a firm's ability to maintain its competitive advantage, despite changes in selected input prices. Project internal flexibility may relate to the sequence of design or construction activities, what resources are used in the design, what tools are applied and other internal issues.

Most implementations of lean philosophy in Norway to date have been in the construction industry and had an internal perspective, as they mainly concerned when, how and by whom the required construction activities were to be carried out. Typically, the design and specifications have not been questioned. This means that currently lean is applied only partially, including tools and techniques, but with a supposed barrier to conversations about value to end user and/or client.

Project external flexibility originates from an effectiveness perspective – to increase the value of the project to the project owner. External flexibility includes questioning the design – Is this really what creates most value for the project owner? Lean focuses on value for the customer who consumes a product or service (Womack *et al.*, 1990).

Based on lean principles, agile software development practices seek to eliminate waste in IT projects (Ågerfalk and Fitzgerald, 2006). In particular, agile methods embrace continuous customer collaboration to assist the development of a gradually evolving IT product (Cohen *et al.*, 2004). Agile methods in IT projects (see e.g. Dingsøyr *et al.*, 2010) include lean applications of both internal and external flexibility. In cooperation with the project owner and user representatives, continual prioritization is performed regarding what functionality is implemented.

LEAN THINKING IN DESIGN

The software development industry relatively early adopted the concept of lean thinking to meet client and market expectations (Poppendieck, 2002). Most of the characteristics of software development are relevant to the construction industry and design development. Work on usability, originally related to product design and software industry, has in a similar way been adapted to buildings (Alexander *et al.*, 2005). According to the ISO standard on usability, a product should be efficient, effective and satisfactory for a given user (ISO 9241-11, 1998). Lean thinking in design has at least two main objectives: to find the best design to meet the clients' needs in order to support effectiveness, efficiency and user satisfaction; and to define systems, structures and materials to ensure effective, streamlined construction. Design management ensures that these objectives are met.

From a traditional perspective, iterations are seen as wasteful when they are required to make corrections, whereas linear processes are regarded as positive in terms of improving productivity. However, if design iterations lead to improved customer value, they are not considered wasteful. Ballard (2000) discusses the important distinction between the two archetypes of activities of making and designing, where designing an artefact requires an understanding and interpretation of the purpose. Making an artefact involves following up the requirements and descriptions expressed in the design in anticipation that the design will realize the purpose. Adapting design to lean thinking requires a fundamentally different approach. Where iterations create value in improving the design, iterations in production, such as rework, imply waste. The design process is characterized by the translation and transformation of ideas, expectations and user needs to find and evaluate multiple solutions to a problem (Peña, 1987). Iterations are necessary to be innovative and find adequate concepts and designs. A first step in lean thinking is to understand fully what value is for the specific task or project and to identify what activities and resources are absolutely necessary to create that value (Poppendieck and Poppendieck, 2007).

Emmitt *et al.* (2004) argue that moving lean thinking upstream – starting with the briefing and conceptual design stages and managing the flow of decisions through to the completed building, creating a greater synergy between design, manufacturing and construction – should create significant potential to deliver value throughout the whole process. In our view, the ultimate consequence of lean thinking in design is the usability of the completed building and how the building supports the core business. To support this, an adequate management regime is required to enhance the different activities and processes, ensuring the best possible outcome within the project's framework.

FLEXIBILITY IN DESIGN PROCESSES: A LAYERED OR STEP-BY-STEP APPROACH

The concept of project flexibility highlights a dual perspective regarding projects. According to Blyth and Worthington (2010), it is normal to find some very strong contradictory conditions in projects. One such contradiction involves alternative perspectives on projects when two parties are involved. For instance, the client and users (as one party) will focus on organizational needs and requirements, whereas the contractor will focus on the technical aspects of the building in question. However, the most important dilemma in a building project concerns the conflicting interests between the demand and supply side regarding the decision-making process. The demand side of a project will focus on the project's effectiveness and aim to keep options open as long as possible or necessary. The supply side will often aim to lock the process early in order to control time and costs.

According to Eikeland (2001) and Olsson (2006), a decision is within room for manoeuvring if it does not violate the consequences of previous decisions. Figure 1 shows that in the early phase of projects (area A), there is larger actual room for manoeuvring than what is typically utilized. This is not the case

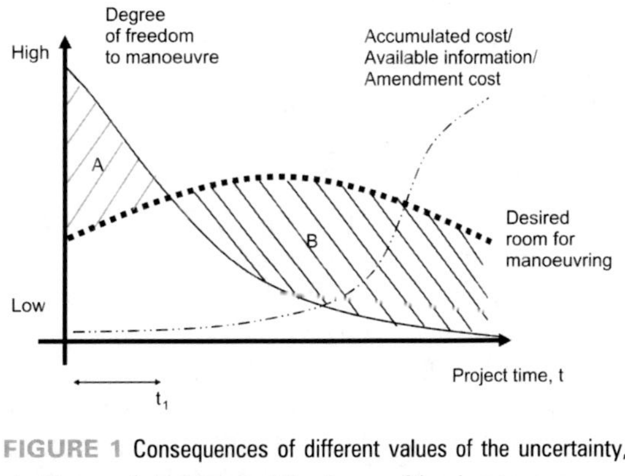

FIGURE 1 Consequences of different values of the uncertainty, significance of decisions and the degree of freedom to manoeuvre compared with the desired room for manoeuvring in different project phases (based on Eikeland, 2001, p.40)

during later parts of projects (area B) where the desired room for manoeuvring is usually limited. Olsson (2006) has provided empirical data that support the model shown. Eikeland (2001) pointed out that a major challenge in project management is that the need for room for manoeuvring is typically at its highest when the freedom for manoeuvring has already decreased significantly. Area B in Figure 1 represents situations where some stakeholders' (e.g. users or clients) desire for room for manoeuvring is greater than the room available. To satisfy the desire for adjustments, changes have to be made because the adjustments violate previous decisions. In Figure 1, one objective of lean design is to utilize and increase area A.

Having knowledge of when and what kind of decisions have to be made is crucial to the success of any project. A project management approach to this should therefore include a strategy for fixing decisions progressively throughout the project, enabling the client to keep options open regarding matters that do not need to be decided or cannot be decided early (Blyth and Worthington, 2010). This leads to a step-by-step approach or a concept with layered decisions. This approach is not new in project management. Decisions tend to be layered from more strategic and general decisions in the initial phases of a project to more detailed focus in the construction and completion phases. Blyth and Worthington (2010) argue that the concept of layering is important because it indicates when certain types of decisions should be made and by whom.

ADAPTABILITY IN BUILDING DESIGN: THE CONCEPT OF LAYERED MODELS

Flexibility in a project can be associated with the decision process or the final product (Gill *et al.*, 2005; Olsson, 2006). Flexibility in the decision process is based on an approach where commitments in projects are made sequentially. This philosophy is developed further by combining the concept of layered building designs (Brand, 1994) and adaptability to a layered decision process (Blakstad, 2001; Arge and Blakstad, 2010). A building can be subdivided into several layers according to the different subsystems' functions and lifespan, from a long-term perspective on main structures to a shorter-term perspective on layout and settings. Combined with a strategy for construction, the design and decisions relating to the different layers can be related to production on site. Arge and Blakstad (2010) state that in order to have an open and flexible process, the different layers in a model should be designed with minimum specifications in order to leave some room to manoeuvre, to accommodate changes at a later date. The individual layers should ideally not be too integrated or connected to each other, and the design should reduce the dependencies between the different layers. This approach is in line with the literature on open buildings, also referred to in publications on lean construction (Koskela, 2000). Habraken (1972) introduced different levels of decision making in the building process in order to accommodate unforeseen changes.

Flexibility can be related to modularity. Modularity refers to the possibility of dividing a project into more or less independent sub-units. Design modularity is a common approach to achieve flexibility by decoupling physical and functional components (Thomke, 1997; Hellström and Wikström, 2005).

Flexibility in the product is related to physical characteristics of a building or other infrastructure. Flexibility as a physical characteristic implies that the building is prepared for possible future changes in use or requirements. If requirements are altered, few changes may be needed because the design can accommodate the revised requirements. According to Pati *et al.* (2008), three aspects of this type of flexibility are adaptability, convertibility and expandability. A similar classification is used by Arge and Landstad (2002) and Bjørberg and Verweij (2009). Adaptability can be defined as a building's ability to meet shifting demands without physical changes. Convertibility can be defined as the possibility for construction and technical changes with minimum cost and disturbance. The third type of flexibility, expandability, can be defined as the ability to increase (or reduce) the size of the building. An alternative term is elasticity. In this article, all three characteristics are seen as subsets of flexibility.

FLEXIBLE PROCESSES AND PRODUCTS

The two dimensions of flexibility in the design process and in the product can be applied independently and also in combination. Such applications have been analysed by, for example, Brand (1994), Slaughter (2001), Boehm and Turner (2003) and Gill *et al.* (2005). Flexibility in the design process and in the product may interact in any given project. Different types of interaction are depicted in Figure 2, each characterized by high or low flexibility in the process and product, respectively. High flexibility in the design process includes changes, extensions and iterations in the project preparations.

The situation with low flexibility in both the product and the design process means that no preparations for project flexibility are made. A strategy characterized by high flexibility in the product and low flexibility in the process is termed a 'robust concept'. A basic principle in the situation with low flexibility in the product and high flexibility in the process is that final decisions can be postponed in order to gain as much knowledge as possible. Such reasoning is also found in real options (see e.g. Amram and Kulatlaka, 1999; Brennan and Trigeorgis, 2000). In real options, flexibility is seen as owning an option – the right, but not the obligation – to take an action in the future (Amram and Kulatlaka, 1999). The value of flexibility can be quantified in monetary terms. Uncertainty can increase the value of a project as long as flexibility is preserved and resources are not irreversibly committed. Uncertainty about the future profitability of a project means that it is often beneficial to postpone commitment to a project until the last responsible moment (Brennan and Trigeorgis, 2000). Set-based design (Ward *et al.*, 1995) can be seen as a real options approach in design by generating different design alternatives. However, a real options approach focuses more on quantification of the value of flexibility.

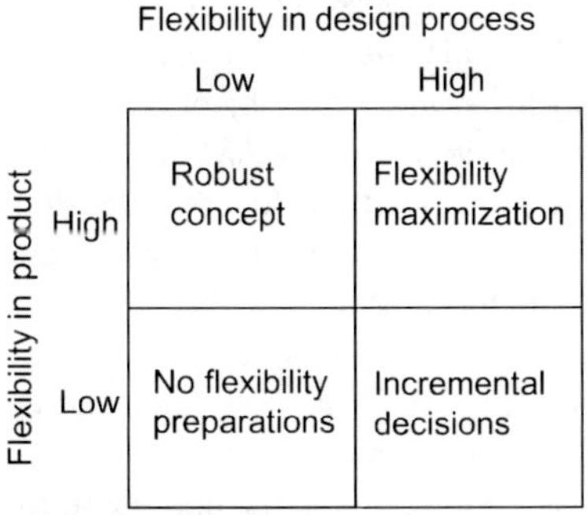

FIGURE 2 Flexibility in the product and design process
***Source:* Based on Olsson, 2006.**

Galbraith (1973) discusses information processing in organizations related to uncertainty. One strategy for reducing the need for information processing is the creation of self-contained tasks, which Galbraith discusses in general terms. Self-contained tasks may be achieved by modularization of the design, by use of standardized components or clarified interfaces related to both design and organization. According to Miller and Lessard (2000), modularity can enable projects to cope with uncertainty because individual components do not have a critical role. An approach of minimal commitment at each decision stage is part of the 'anti-disaster methodology' proposed by Hall (1980, p.267). In essence, Hall (1980) proposes that decisions should be made at the last responsible moment, although his work predated the use of this term.

With further regard to Figure 2, Olsson (2006) found indications that projects with highly flexible product designs also had flexible decision processes as a response to contextual uncertainty. If this is the case, flexible products and flexible decision processes cannot compensate for each other in a clear-cut way as described in relation to Figure 2.

METHOD AND MATERIALS

This article is based on a study of lean and flexibility aspects in two hospital projects in Trondheim and Bern. A qualitative case study research approach was used (Yin, 2003). This study is a case study based on trailing research (Finne *et al.*, 1995).

Separate sets of research material have been used, particularly a combination of personal qualitative experience and interviews. The most important sources of information were:

- project documentation and publicly available information;
- participant observation during project planning and design;
- interviews with key personnel.

The authors collected all important information related to the investigation in a case-study database. The database included sources of information, collected documentation and interview guidelines. Both the hospital projects are well described in evaluation reports, which were used as supporting documentation and for quality assurance of the presented facts.

Multiple sources were used as a basis for the conclusions drawn. Further, as described above, we used input from multiple respondents in each project to reduce the possibility of bias in the conclusions. On the basis of the case studies, an assessment was made regarding the characteristics of layered design and layered processes in the respective projects.

One difference between the two cases concerns the fact that one of the authors was involved as an adviser and evaluator in different project phases in one of the cases, the St Olavs case. We therefore had access to project documentation and were able to participate in meetings and other project activities. During quality assurance of this case, special attention was paid to investigating whether the case description was subject to personal bias on behalf of the researchers. The review largely supported the researcher's understanding of the process, but some clarifications were made.

TWO CASES FOR ILLUSTRATION

As described by de Neufville *et al.* (2008) and Miller and Swensson (2002), flexibility is an important concern and presents special challenges in hospital buildings. Due to changes in medical technology, regulations, organization and funding of health-care services, demography and epidemiological patterns, demands on hospital buildings are likely to change.

Large hospital construction projects are planned up to 10–20 years in advance. Hospitals are typically designed to have a lifespan of more than 40 years. This creates a need for flexibility and also presents huge challenges regarding effective planning and construction processes. Walker and Shen (2002) studied flexibility in a hospital construction project and point out that the ability to utilize flexibility options is influenced both by the ability of an organization to be flexible and by its commitment to be flexible. Further, de Neufville *et al.* (2008) discuss flexibility in terms of organization, management and the use of facilities. Parrish *et al.* (2008) describe how lean concepts, including set-based design, can be applied in the design phase of a hospital project.

Neither of the two cases described in this article have had a specific focus on lean thinking in the design phase, but are used to illustrate how flexibility in both the design process and the product can be an adequate strategy to meet different internal and external changes during a long-lasting planning and design process, enabling manoeuvring room for right decisions to be made in a long-term perspective. The concept of lean thinking is discussed in this context.

ST OLAVS HOSPITAL, TRONDHEIM

Plans for a new university hospital were prepared in 1991, and approved by the Norwegian Parliament in 1993. The new hospital will ultimately consist of 10 different medical centres (Figure 3). The project was divided into two construction phases. The first phase of the construction, consisting of four centres and totalling 88,000 m^2, was completed in 2006. The total cost of Phase 1 was US$730 million. Phase 2 is scheduled for completion in 2014, by which time the completed hospital will total almost 220,000 m^2 and is expected to reach a total cost of US$1,570 million.

The construction project, together with initiatives for organizational change and improvement, was intended to provide significant gains in terms of productivity, including quality improvements (Myrbostad, 2003).

LAYERED PHYSICAL DESIGN

The development plan for the university hospital in Trondheim contains the vision, objectives and framework conditions for the project. These are expressed as seven main objectives that elaborate upon and finalize the vision. Two of these objectives are especially important for the development of the physical design and have relevance for the discussion in this article:

- *Adaptability*: the project must be adaptable and flexible to incorporate future changes.
- *Efficiency*: the project must contribute to an effective organization and represent value for money.

FIGURE 3 St Olavs Hospital master plan (*Source:* Arch. Team St Olavs)

The new hospital in Trondheim was originally planned according to a clinical centre model. Capacities and functions in each of the six Clinical Centres were planned based on ICD-10 diagnostic groups. To a large extent, each centre was intended to be self-supporting in terms of clinical support resources and competence.

The design has been developed from an overall perspective with a master plan for the whole hospital, to design principles for each centre, design principles for technical infrastructure and systems, and the layout and design of technical equipment. An important strategy in the planning and design of the hospital was the concept of a Generic Centre as a virtual centre, describing general and fundamental solutions common for all Clinical Centres in the hospital project. The idea behind the Generic Centre was to establish a prototype for the design, and the final design of the individual centres (Figure 4) comprised minor variations of this common 'design platform' named the Generic Centre.

Through standardization, the project has sought to achieve better and more efficient planning and operation. The concept comprises guidelines and instructions regarding basic principles for building structure, communication and logistics, standard room categories, infrastructure and architecture, to ensure a common overall structure, character and recognition. In addition, there has been a principle of having extra capacity in the foundation and main constructions to allow for a 20% increase in technical installations and have a space reserve of 25%. Some of the space reserves have already been used due to changes and increases in the space programmed during the planning and design. The

FIGURE 4 One of the Clinical Centres – The Neuro Centre
Source: Arch. Frisk Arkitekter AS; Photo: G. K. Hansen, 2009

design concept for the hospital has been robust and the changes have not had any major negative consequences.

LAYERED DESIGN PROCESS

Flexibility in the decision and planning process has been an important issue in the project. A flexible process allows for adjustments throughout the project, and also relates to the ongoing operations of the existing hospital. The planning of the project is based on a step-by-step process, with planning and decision making from an overall level to a more detailed and specific level. The process is a layered and iterative process with alternation between process analyses, programming and designing with different levels of detail. The briefing process was divided into three main levels: the main function brief (level 1); a building project brief for each of the Clinical Centres (level 2); and a space management brief for each centre (level 3). In addition, a technical brief for the project was developed.

A project such as St Olavs Hospital has a need for specific and detailed knowledge of current activities and capacities, as well as possible avenues towards improving conditions for the future. Obtaining such knowledge requires the involvement of employees and specialists from different levels and places in the organization. This argues for a bottom-up approach to create an enterprise map and identify areas for improvement. However, there is also a need to make decisions and priorities that trigger the relevant changes and ensure that the project is working towards common goals, implying the need for elements of a top-down approach.

Many of the discussions around the development of the architectural design regarding the use of space, functionality and adaptability were related to the Generic Centre. There were some disagreements about the principles described, and whether they should be absolute or guiding. Regarding the planning of the different centres, the users wanted individual solutions and not 'standard' solutions developed from the generic centre concept (Paulsen *et al.*, 2003).

INSELSPITAL, HOSPITAL DE L' ILE, BERN

The INO project at Insel University Hospital in Bern (Figure 6) is a new intensive and acute care and surgical unit in Bern. The size of the project is *c.* 54,000 m^2 (gross). The first phase of the project has been completed and has an area of 25,400 m^2. The next phase, consisting of 25,000 m^2, is currently under construction. The site for the new project lies within the existing hospital campus. Parts of the existing hospital have been demolished, and the location serves as the site for the new INO unit. The focus on change and development has been important from the beginning of the project.

The relationships between the operating budget, business planning, and the construction process and construction costs have been central to the strategy. The analyses and description of the business

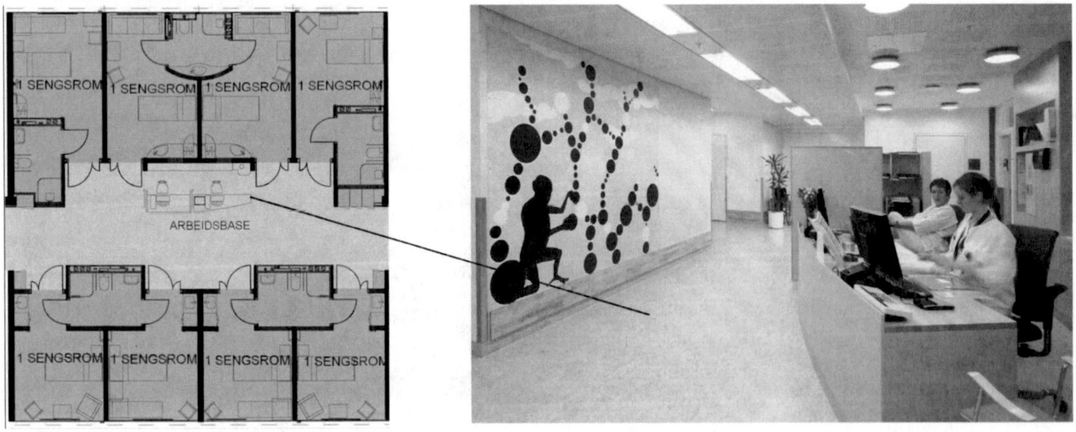

FIGURE 5 Bed cluster and open workstation (*Source:* Arch. Team St Olavs/photo: Grethe Britt Fredriksen)

FIGURE 6 Inselspital, Bern, facade
Source: Photo: G. K. Hansen, 2010.

development are based on three levels or aspects, and form the basis for further planning and design. *The process concept* establishes the long-term operational perspective (15–50 years), in which capacity, development potential, flexibility, space management and the main layout are determined. The process concept forms the framework for *the organization concept*. The organization concept has a time horizon of 5–15 years and describes the organization of the hospital, the medical–technical concept and the functional requirements, which form the basis for the design process. *The operational concept* defines the short-term solutions and has a time horizon of 0–5 years. This concept primarily deals with space management, fixtures and equipment, and how daily operations are organized. The impact on and consequences for costs, priorities and risk have been assessed for all three concepts.

LAYERED PHYSICAL DESIGN

The INO project planned a process where the building is regarded as consisting of three systems. The *primary system* is a long-term investment and is regarded as fixed with a time horizon of 100 years (Figure 7). The primary system's base building is expected to accommodate changing departmental sizes and changing layouts of the emergency, imaging, surgery and pharmacy departments, zoned for specific floors (the building was not designed to have nursing units). There are three vertical cores and/or shafts in the corners of the building. Apart

FIGURE 7 Inselspital, Bern, floor plan (*Source:* Kamm + Kundig Architects)

from structural components and vertical shafts, the floor plans are completely open. This provides a high degree of generality, where walls, furnishings and equipment can be placed where desired.

The secondary system is adaptable and has a shorter time horizon in terms of investment, and is intended to function for more than 20 years. The secondary system deals with the measures needed to realize organizational concept, and mainly concerns the plan principles, organization of departments, principles and systems of technical infrastructure, lighting, communications and supply, and fire and safety. The secondary system has to be adapted to the primary system, and at the same time be maximally independent of the primary system due to its shorter functional lifetime.

The tertiary system should be flexible and is linked to short-term investment where changes can occur without significant structural implications. The tertiary system includes fixtures, fixed or mobile medical technical equipment, and an interior design with specified requirements, and is expected to have a useful life of 5–10 years or an even shorter perspective. It is emphasized that the tertiary system should be independent of primary and secondary systems as far as possible and that changes within the system should be made with as few consequences as possible for the other systems.

LAYERED DESIGN PROCESS

The INO project has adopted an entirely new process for procuring the facility. A hierarchical structure of decision levels was developed as a management model. Higher-level decisions provide 'capacity' for a range of lower-level functional scenarios. The principle is to partition the whole by what is called 'systems separation'. One architect designs the base building, without detailed programmatic information about what will be filled in later.

A competition was held to select a design and construction firm for each of three distinct 'system levels' as a step-by-step process, where subsequent competitions were based on previous results. In principle, there were three different planning teams for the three different systems or levels.

The first competition was limited to a master plan for maximizing the development and use of the site and an overall concept for main building structures (won by Peter Kamm + Kundig Architects). This was followed by a second competition for the selection of the secondary system architect, to engage in intense programming with the client's hospital staff and subsequently propose the secondary system to be installed in the base building (won by Itten Brechbühl Architects). This strict separation was made to ensure that the base building is not designed for just one programmatic scenario. The third and final competition was held to select the firm to provide the tertiary system (won by HWP Architects). In this process, the separate contracts were organized to reduce the typical dependencies among the three system levels.

The approach yielded benefits even before Phase I was completed. Functional layouts had already morphed because of changes in the organization's priorities and client base, as well as medical procedures, hospital politics and technology.

DISCUSSION

Thus far we have described different aspects of flexibility and how they can contribute to the analysis and management of project flexibility. According to Olsson (2006), lean and flexibility aspects can be discussed with reference to several dimensions, such as flexibility perspectives, and flexibility drivers and enablers.

DRIVERS

In this article, we use the term 'drivers' as factors that create needs or pressure on projects to be flexible. Uncertainty and risk appear to be the key drivers for project flexibility, along with a need for increased productivity. Uncertainty can occur as a gap between the amount of information needed to make a decision and the amount of information available. In the studied projects, the information gap has been addressed by aiming to both reduce the amount of information needed and increase the available amount of information. A long duration of projects is likely to result in a need for scope changes. In the case of the INO project, the focus on change has been important from the outset, and the driver for a new approach towards 'change-ready' hospital designs. The St Olavs Hospital project experienced a significant change in the framework conditions during the project period, as the responsibility for the hospital sector was transferred from county level to national level.

One aspect of the effectiveness of a hospital building project is the cost and quality level of the operations in the new building (i.e. the processes within a new hospital). The same cost and quality measures will be related to the efficiency of the hospital when the new buildings are taken into use. In other words, the efficiency in the hospital (once it is built and taken into use) is one dimension of effectiveness of the construction. An overall objective of user involvement at St Olavs was effectiveness in terms of suitable future hospital facilities. Value maximizing will therefore be more important than waste reduction in the long term, but the two dimensions of lean thinking cannot be seen in isolation or in contradiction to one another. Emmitt *et al.* (2004) argue that moving lean thinking upstream will create a greater synergy between design, manufacturing and construction, thereby delivering value throughout the whole process. Lean thinking will, of course, support improved processes and reduce waste, but more importantly it will lead to value for money for all stakeholders in a project. Recent studies of the usability of workplaces underline the importance of focusing on the effect of the building with a long-term perspective on the use and operation of the building (Alexander *et al.*, 2005; Blakstad *et al.*, 2008). The two examples from the hospital projects are in line with this philosophy, not only enabling a streamlined process and changes during design and construction, but also opening up for future demands and changes with as few costs and interruptions as possible for the core activities.

FLEXIBILITY ENABLERS

In this article, we refer to 'enablers' as factors that contribute to making it possible for projects to be flexible in terms of having room for manoeuvre, that is, the possibility to delay decisions until the last responsible moment and open up for 'right in time decisions' rather than lock the process too early. We have focused on the dimensions of flexible products and processes as possible strategies in lean thinking in design.

According to Olsson (2006), project flexibility can be divided into flexibility in the design process and the product. The St Olavs project can be described as having relatively high flexibility in the product but medium flexibility regarding the decision process; thus the concept can be defined as quite robust with a certain amount of room for manoeuvring. The concept for the INO project can be described as flexibility maximization, with a high degree of flexibility in both the product and the process.

With regard to the planning and design process, the two cases have chosen different strategies but with the same intentions of enlarging the room for manoeuvring. The St Olavs project had a clear aim of flexibility in both the planning process and the physical model. The process at St Olavs is a combination of an iterative process between the programming and design process, and a step-by-step approach from an overall perspective to more detailed planning. The iterative process, not only between the different levels in programming or design but also between programming and design, should ideally result in improved learning and understanding of the demands (in programming) and the possibilities (in design). This is in accordance with, for example, Amram and Kulatlaka (1999). The development of the design was also planned to be 'mirrored' by an intertwined process regarding the organizational concept of the hospital.

The INO project had a step-by-step process with clear definitions of scopes and requirements for each of the three systems, and seems to be more linear. An important objective in this respect was to establish a loose-fit connection between the base building (defined in terms of accommodation capacity) and the structural, spatial and equipment needs of the fit-out. In this process, modularity, as described by Miller and Lessard (2000), served as an enabler for flexible project management.

The INO project is based on a philosophy of leaving maximum decision flexibility to designers at the next level, even if one design firm is responsible for designing all levels. Compared with the St Olavs project, the INO project seems to be more expert driven regarding the design process. An interesting aspect of the INO project is the strong focus on lifespan, regarding not only the physical model, but also the analyses of the process concept, the organization concept and the operational concept, the latter having a time horizon of 0–5 years. This also reflects and influences the driving forces for change to the physical and technical systems of the hospital.

The two cases are also interesting because they illustrate different strategies regarding the design concepts. The St Olavs project approach to the design concept is planning for adaptability. Here, reserve capacity on physical structures combined with standardization regarding functionality (room typologies), equipment, technical systems and solutions have been important strategies. The concept of the Generic Centre as a virtual model for planning, designing and decisions is interesting as a potential flexibility enabler. To some extent, the adoption of this concept has parallels with the use of platforms in the automobile industry, ensuring high quality and standardization on the most basic and long-lasting parts of the product, and allowing adaption and specification for different clients and markets.

The philosophy for the INO project is that the different systems and elements in a building have and should be planned for different lifetimes or time horizons, with each system seen as independent and as open as possible to the other systems. This is quite in line with Brand (1994) and literature on open-building concepts (Habraken, 1972). There are plans for reserve capacity regarding floor plans and vertical shafts, and also load capacity on floors. The INO project represents a very deliberate strategy with regard to the planning process and in relation to the physical model. The strategy focuses on both structural and cultural components in relation to organizational development and the quality of the input that is given. Experiences to date with this model have revealed the challenges regarding the division of the design and construction processes. Defining the interfaces between system levels was more important and difficult to accomplish than was initially anticipated.

CONCLUSIONS

Both St Olavs Hospital and the INO project combined several elements to increase their room for manoeuvring, combining flexibility in both product and processes. In a project management perspective, lean thinking means streamlining the overall process to handle changes and open up for right decisions at the right level and at the right time, with as few consequences or interferences to existing or future decisions as possible. In this article, we have demonstrated the importance of seeing the design process and design model as two integrated parts that can either support or hinder the outcome of a project.

The two cases demonstrate the importance of a strong leadership, a clear vision and a long-term perspective on the desired outcome of the project. From a management perspective, lean thinking will be useful in discussing and identifying when the different processes produce waste or increase value. One of the challenges in the two cases presented here has been the interfaces between the different processes, levels and actors. These have to be improved to ensure both an effective process and a product giving the right effect.

Traditionally, lean thinking in production has had an internal focus. This article has underlined the importance of recognizing the design process as a significant process other than production and one that needs a separate approach. The design is a more open and iterative process, where changes and iterations are vital to improve the design and the final product. Lean thinking is primarily designed to reduce waste and increase value for the different actors in the process. To evaluate the design, a clear framework has to be developed. Traditionally, the design has been evaluated against the programme or the brief, to identify which technical and functional aspects of the design are to be met. Lean thinking regarding increased value requires a clear understanding and description of the customer's values, in which the design is seen as one instrument to achieve this. The two cases examined in this article show different approaches to dealing with this, but also reveal the need for a stronger focus on how to evaluate and choose the design alternatives giving the best value.

Recently, some interesting discussions have emerged on bringing together the concepts of lean thinking, value management and design management in order to enhance the value of both the process and the outcome.

REFERENCES

Ågerfalk, P.J. and Fitzgerald, B., 2006, 'Flexible and distributed software processes: old petunias in new bowls?', *Communications of the ACM* **49**, 27–34.

Alexander, K.,, Fenker, M., Granath, J.Å., Haugen, T. and Nissinen, K., 2005, 'Usable workplaces: action research', in: *Proceedings. CiB 2005, Combining Forces – Advanced Facilities Management & Construction through Innovation Series*, Helsinki, Finland, 13–16 June 2005, 389–399.

Amram, M. and Kulatlaka, N., 1999, *Real Options: Managing Strategic Investment in an Uncertain World*, Harvard Business School Press, Financial Management Association Survey and Synthesis Series, Boston, MA,

Arge, K. and Blakstad, S. H. 2010, 'Briefing for adaptability', in: A. Blyth and J. Worthington (eds), *Managing the Brief for Better Design*, Routledge, London and New York, 159–170.

Arge, K. and Landstad, K., 2002, 'Generalitet, fleksibilitet og elastisitet i bygninger. Prinsipper og egenskaper som gir tilpasningsdyktige kontorbygninger [Generality, flexibility and elasticity in buildings. Principles for adaptability of office buildings.]', Norges byggforskningsinstitutt Prosjektrapport 336, Norges byggforskningsinstitutt, Oslo.

Ballard, G., 2000, 'Positive vs. negative iteration in design', in: *Proceedings of the 8th Annual Conference of the International Group for Lean Construction (IGLC-8)*, University of Sussex, Brighton, 44–55.

Ballard, G. and Howell, G.A., 2003, 'Lean project management', *Building Research & Information* **31**(2), 119–133.

Bjørberg, S. and Verweij, M., 2009, 'Life-cycle economics: cost, functionality and adaptability', in: B. Rechel, S. Wright, N. Edwards, B. Dowdeswell and M. McKee (eds), *Investing in Hospitals of the Future*, WHO Regional Office for Europe, Copenhagen, 145–166.

Blakstad, S.H., 2001, 'A strategic approach to adaptability in office buildings', Doctoral dissertation No. 2001:97, Norwegian University of Science and Technology, Trondheim.

Blakstad, S.H., Hansen, G.K. and Knudsen, W., 2008. 'Methods & tools for evaluation of usability in buildings', in: K. Alexander (ed), *Usability of Workplaces, Phase 2*, CIB W111 Research Report, International Council for Research and Innovation in Building and Construction, Rotterdam.

Blyth, A. and Worthington, J., 2010, *Managing the Brief for Better Design*, Routledge, London and New York.

Boehm, B. and Turner, R., 2003, 'Observations on balancing discipline and agility', in: *Proceedings of the Agile Development Conference ADC'03*. IEEE Computer Society, Salt Lake City, UT, 25–28 June.

Brand, S., 1994, *How Buildings Learn: What Happens After They're Built*, Penguin Books, New York.

Brennan, M.L. and Trigeorgis, L., 2000, *Project Flexibility, Agency, and Competition: New Developments in the Theory and Application of Real Options*, Oxford University Press, New York.

Cohen, D., Lindvall, M. and Costa, P., 2004, 'An introduction to agile methods', *Journal of Advances in Computers* **26**, 1–63.

Cremieux, P.-Y., Ouellette, P., Rimbaud, F. and Vigeant, S., 2005, 'Hospital cost flexibility in the presence of many outputs: a public–private comparison', *Health Care Management Science* **8**, 111–120.

de Neufville, R., Lee, Y.S. and Scholtes, S., 2008, 'Flexibility in hospital infrastructure design', in: *IEEE Conference on Infrastructure Systems*, Rotterdam, 10–12 November 2008.

Dingsøyr, T., Dybå, T. and Brede Moe, N. (eds), 2010, *Agile Software Development*, Springer, Berlin, Heidelberg.

Eikeland, P.T., 2001, *Teoretisk Analyse av Byggeprosesser [Theoretical Analysis of Construction Processes]*, Samspill i byggeprosessen, prosjektnr. 10602, NTNU, Trondheim.

Emmitt, S., Sander, D. and Christoffersen, A.K., 2004, 'Implementing value through lean design management', in: *Proceedings IGLC2004, Copenhagen*. Available at www.iglc2004.dk [accessed November 2010].

Finne, H., Levin, M. and Nilsen, T., 1995, 'Trailing research: a model for useful program evaluation', IFIM-paper 4/95, SINTEF IFIM, Trondheim.

Galbraith, J.R., 1973, *Designing Complex Organizations*, Addison-Wesley, Reading, MA.

Gill, N., Tommelein, I.D., Stout, A. and Garrett, T., 2005, 'Embodying product and process flexibility to cope with challenging project deliveries', *Journal of Construction Engineering and Management* **131**(4), 439–448.

Habraken, N.J., 1972, *Supports: An Alternative to Mass Housing*, Architectural Press, London.

Hall, P., 1980, *Great Planning Disasters*, Weidenfeld and Nicolson, London.

Hellström, M. and Wikström, K., 2005, 'Project business concepts based on modularity – improved manoeuvrability through unstable structures', *International Journal of Project Management* **23**(5), 392–397.

ISO 9241-11, 1998, *Ergonomic Requirements for Office Work with Visual Display Terminals (VDTs) – Part 11: Guidance on Usability*, International Organization for Standardization, Geneva.

Koskela, L., 1992, 'Application of the new production philosophy to construction', CIFE Technical Report 72, CIFE, Stanford University, Stanford, CA.

Koskela, L., 2000, 'An exploration towards a production theory and its application to construction', PhD thesis, VTT Publications 408, VTT, Technical Research Centre of Finland, Espoo.

Mikkelsen, H. and Riis, J.O., 2003, *Grunnbog i prosjektledelse [Textbook of Project Management]*, Prodevo ApS, Rungsted.

Miller, R. and Lessard, D.R., 2000, *The Strategic Management of Large Engineering Projects: Shaping Institutions, Risks and Governance*, MIT Press, Cambridge, MA.

Miller, R.L. and Swensson, E.S., 2002, *Hospital and Healthcare Facility Design*, 2nd edn, W.W. Norton, New York.

Myrbostad, A., 2003, 'Bruk av prosessanalyser i sykehusplanlegging', HMT-Helse/Medisin/Teknikk Nr 5–2003.

Nahm, Y.-E. and Ishikawa, H., 2005, 'Representing and aggregating engineering quantities with preference structure for set-based concurrent engineering', *Concurrent Engineering* **13**(2), 123–133.

OECD, 2000, *OECD Glossary of Terms in Evaluation and Results Based Management*, Organisation for Economic Co-operation and Development, Paris.

Olsson, N.O.E., 2006, 'Management of flexibility in projects', *International Journal of Project Management* **24**(1), 66–74.

Olsson, N.O.E., 2008, 'External and internal flexibility – aligning projects with the business strategy and executing projects efficiently', *International Journal of Project Organisation and Management* **1**(1), 47.

Parrish, K., Wong, J.M., Tommelein, I.D. and Stojadinovic, B., 2008, 'Set-based design: case study on innovative hospital design', in: P. Tzortzopoulos and M. Kagioglou (eds), *Proceedings of the 16th Annual Conference of the International Group for Lean Construction*, Manchester, UK, 16–18 July.

Pati, D., Harvey, T. and Cason, C., 2008, 'Inpatient unit flexibility design characteristics of a successful flexible unit', *Environment and Behavior* **40**(2), 205–232.

Paulsen, B., Berg, T., Jordanger, I. and Hansen, G.K. 2003, 'Plan og utviklingsprosessen i Helsebygg Midt-Norge', SINTEF Rapport STF22 A03502, SINTEF, Trondheim.

Peña, W., 1987, *Problem Seeking: An Architectural Programming Primer*, AIA Press, Washington, DC.

Poppendieck, M., 2002, *Principles of Lean Thinking* [available at www.poppendieck.com/papers/LeanThinking.pdf] (accessed 30 October 2010).

Poppendieck, M. and Poppendieck, T., 2003, *Lean Software Development: An Agile Toolkit*, Addison-Wesley, Canada.

Poppendieck, M. and Poppendieck, T., 2007, *Implementing Lean Software Development: From Concept to Cash*, Addison-Wesley, Canada.

Slaughter, E.S., 2001, 'Design strategies to increase building flexibility', *Building Research and Information* **29**(3), 208–217.

Sobek, D.K., Ward, A.C. and Liker, J.K., 1999, 'Toyota's principles of set-based concurrent engineering', *Sloan Management Review* **40**(2), 67–83.

Thomke, S.H., 1997, 'The role of flexibility in the development of new products: an empirical study', *Research Policy* **26**(1), 105–119.

Tommelein, I.D., 1998, 'Pull-driven scheduling for pipe-spool installation: simulation of lean construction technique', *ASCE Journal of Construction Engineering and Management* **124**(4), 279–288.

Walker, D.H.T. and Shen, Y.J., 2002, 'Project understanding, planning, flexibility of management action and construction time performance: two Australian case studies', *Construction Management and Economics* **20**(1), 31–44.

Ward, A., Liker, J.K., Cristiano, J.J. and Sobek, D.K., 1995, 'The second Toyota paradox: how delaying decisions can make better cars faster', *Sloan Management Review* **36**(3), 43–61.

Winch, G.M., 2010, *Managing Construction Projects*, John Wiley & Sons, Chichester.

Womack, J.P., Jones, D.T. and Roos, D., 1990, *The Machine that Changed the World*, Rawson Associates, New York.

Yin, R.K., 2003, *Case Study Research: Design and Methods*, 3rd edn, Sage, Thousand Oaks, CA.

An Empirical Study of Information Flows in Multidisciplinary Civil Engineering Design Teams using Lean Measures

Effi Tribelsky[1] and Rafael Sacks[2]

[1]Israel Air Force
[2]Faculty of Civil and Environmental Engineering, 840 Rabin Building, Technion Campus, Haifa 32000, Israel

Abstract

Achieving smooth and continual flow of information among teams of independent designers is one of the central difficulties in detailed design of complex civil engineering projects. Lean production flow phenomena such as bottlenecks, rework, large batches and long cycle times are common. This research sought correlation between the degree of success achieved, in terms of design document effectiveness, meeting budget and schedule targets, and participant satisfaction, on the one hand, and the occurrence of information flow problems, on the other. Extensive data describing the communication of design documents between team members were collected in the detailed design phase of 14 civil engineering projects, all of which were part of a major airport construction project. The degree of effectiveness of the design documents, the extent to which budget and schedule targets were met, and participants' evaluations of their success were measured. The results confirm positive correlation between the quality of information flows and the effectiveness of design documents, and confirm that unstable information flows are associated with unpredictable project outcomes. No correlation was found between information flow problems and design managers' subjective views of the processes, nor between their performance evaluations and their fulfilment of budget or schedule targets.

INTRODUCTION

Building and infrastructure projects require teams of designers with a broad variety of engineering, architectural and interdisciplinary skills. The design teams are composed ad-hoc for individual projects, the designers belong to multiple independent firms, and they must produce a single coherent design through collaborative effort. The resulting difficulty in sharing information is exacerbated by the fact that the designers do not co-locate for the project, and tend to shift focus from project to project within their own design offices. While the behaviour of commercially independent design consultants within collaborative teams is generally governed by the design contracts signed by each party with the client, the day-to-day functioning of such teams rests heavily on informal relationships and tacit understandings between the team members (Love *et al.*, 2002; Ford and Sterman, 2003).

Designers use and produce information. Ideas, principles, rules and skills are applied to generate an abstract description (a digital building model and/or a set of drawings and specifications) that defines how a physical facility is to be constructed. Information is also designers' raw material. Information must flow between designers, and that flow is a critical determinant of their ability to perform their work effectively and efficiently (Prasad *et al.*, 1993; Moreau and Back, 2000; Eckert *et al.*, 2001; Gray and Hughes, 2001). Inefficient flows of information result in numerous forms of waste:

- Rework follows wherever design has proceeded on the basis of outdated versions of other designers' drawings because newer information was not forthcoming (Ashford, 1989; Love *et al.*, 2000).
- Waiting for information leads designers to shift their attention to other projects, incurring the waste of renewed 'setup' times for familiarization with what was done earlier when work on a project is resumed.
- Overall project durations are extended and overhead and financing costs grow as local cycles of iteration are delayed or repeated due to negative iteration (Ballard, 1998).
- Over-design, which follows from the desire to avoid the need to deal with additional or changed design requirements that may be communicated only later on in the design process.

The flow of information in dispersed design teams is not well understood, partly due to lack of theory and partly due to lack of measures (Tribelsky and Sacks, 2007), and so it is frequently not managed well. Baldwin *et al.* (1999) found that understanding the process by which information flows between project participants is essential to improvement of design management. Emmitt (2007) called for mapping and understanding information flows as a prerequisite for design management in architectural firms, and highlighted the need for an understanding of waste in this context.

The causal link between the quality of design and design documents and the amount of waste and rework during construction itself is well established (Sacks *et al.*, 2003; Thomas *et al.*, 2004; Tilley, 2005; Emmitt, 2007). The evidence on rework in design and construction suggests that its quantity is inversely related to the quality of information flows: where information is made available to all quickly and effectively, rework is reduced. But can evidence be found to link the nature of information flows in detailed engineering design directly to the quality of design process outcomes? Earlier work (Tribelsky and Sacks, 2010) showed that the phenomena that characterize poor flow in generic production systems, such as large batch sizes, long cycle times, flow bottlenecks, accumulation of work in process (WIP) inventories and others, can be identified in flows of information in the detailed phase of civil engineering design, and that they can be measured. The research goal therefore was to collect empirical evidence of the degree to which design information flows were interrupted (or exhibited the phenomena that indicate poor flow) in a sample set of projects, collect data about the design outcomes of those projects, and identify any possible correlation between information flow problems and design outcomes.

The next section of the article reviews the literature on the nature of information flow in design. The methodology section describes how data on 14 projects, all part of a major airport relocation, were collected and analysed using seven lean measures of information flow. These results were compared with assessments of the project outcomes, in terms of both product and process, allowing conclusions to be drawn concerning the relationships between the characteristics of information flow between design team members and the degree of project success or failure they achieved.

INFORMATION FLOW IN DETAILED DESIGN OF CIVIL ENGINEERING PROJECTS

CONCEPTUAL MODELS

In current civil engineering practice, although communication using building information models is becoming more common as the difficulties of interoperability are tackled (Eastman *et al.*, 2008), design information is still most commonly communicated in documents, whether electronic or paper. The information itself is intangible, and measurement of information flow in a design process must be based on a conceptual model of information flow.

A variety of conceptual models have been proposed. Fyall (2002) suggested analysing information flow in organizations using concepts of fluid mechanics and proposed a measure of turbulence in information flow based on the 'Reynold's number'. Krovi *et al.* (2003) also used the fluid mechanics metaphor, proposing fluid flow dynamics measures such as velocity, viscosity and volatility. Ostergaard and Summers (2007) modelled collaborative design and information flow using concepts of electrical current, including calculation of a resistance value for each task through which information must flow. Ballard (2000) considered the detailed design process to be similar to production processes in as far as inputs are transformed into outputs at various processing stations, making the distinction that whereas in the typical production process machines and workers produce tangible products, in the detailed design process consultants transform requirements and ideas into product design documents. Huovila *et al.* (1997) extended the production process analogy to include conceptualization of the flow of information between activities in addition to the transformation at each workstation. In this analogy, information flows in discrete 'packets' that are carried between designers or from one design activity to another (Koskela *et al.*, 2002). The information packets can be copied without limitation, they can be stored, and they can be aggregated or divided up. The same features of flow, such as bottlenecks (where information is not produced or is withheld by a particular designer), batch size and cycle time, are observed in both design processes and production processes.

The production process flow analogy is widely accepted implicitly, in that numerous researchers and practitioners have advocated and explored application of design management techniques derived from techniques that were developed in production contexts. Examples include the Design Structure Matrix approach (Steward, 1981), the Toyota Product Development System (Morgan and Liker, 2006), Theory of Constraints (Goldratt, 1997) and Lean Design in the context of Lean Construction (Freire and Alarcon, 2002).

Most of the literature recognizes the principal difference between production and design concerning perception of waste: in production, variance among products and rework are considered waste, whereas design relies on iteration of some steps. Design comprises cycles of synthesis and analysis (Alexander, 1964). In design, 'straw' values must often be set for some parameters to enable analysis, on which deeper understanding is gained, which generates value even if the design set proved unworkable. The ability of a design team to identify an optimal conceptual solution early in the process is a determinant of success in terms of the cost/benefit of the process as a whole (Manavazhi and Xunzhi, 2001), because it can reduce the need for 'negative' (unproductive) iteration (Ballard, 1998).

MEASURING DESIGN INFORMATION FLOWS

In earlier work (Tribelsky and Sacks, 2010), the authors developed a specialized technique for measuring information flows in design projects. The technique uses seven numerical indices of flow characteristics to identify and verify the presence and intensity of four indicators of process flow problems inspired by the lean conceptual model: flow bottlenecks, rework, large batch sizes and high levels of WIP (Hopp and Spearman, 1996; Koskela, 2000).

The seven numerical indices measure the rates and patterns of generation of design information, dissemination, batching and other features. They are listed in Table 1 and detailed below.

TABLE 1 Information flow indices

INDEX	DESCRIPTION	SYMBOLS	UNITS
Action rate	The rate at which information is transferred	AR	Actions/ time
Package size	Quantifies the level of detail of information packages	PS	Information attributes
Work in process	The number of available but unused information packages	WIP	Information packages
Batch size	The batch volume of information transferred	BS	Information attributes
Development velocity	The velocity of information development as represented by accumulation of detail	DV	Information attributes/ time
Bottlenecks	Identifies possible bottleneck partners in the process at any given time	BN	
Rework	Quantifies the rework included in information packages	RW	Information attributes

Action rate: a direct measure of the behaviour of members of the design team, individually and as a team. It is the number of transactions recorded per day in the project file-sharing website, averaged over periods of time or over periods between milestones. Transactions include viewing, loading, presentation, deletion, update, attachment, etc. using the project file-sharing website. A team member's current rate can be compared with the corresponding average rate to determine the degree of activity of its owner. In particular, irregular flow can be identified where there are peaks in transaction rate well above the average, indicating accumulation of information by a designer in large batches instead of smoother flow, and/or application of work intermittently to the project rather than continually.

Package size: the information package size index quantifies the level of detail of information contained in packages. It enables evaluation of the rate of increase of detail and of the approximate degree of completion.

Work in process: measures the delay in usage of a project's information packages, that is, unused information inventory. At any given point in time, it indicates the total number of information packages in the file-sharing website that have been uploaded but have not been used up to that time, weighted with the duration for which each has been dormant. Long gaps between upload of information and its use by others represent poor flow.

Batch size: a quantitative and qualitative index intended to describe the volume of information transferred by an individual designer or consultant in a single, short pulse.

Development velocity (DV): the rate of publication of information detail as it is made available to members of the team on the extranet over time. The development velocity index is not the actual rate of information development as it occurs in the design office, because there is a delay between actual design activities and publication of the information. The index uses the publication date, which is later than the creation date for most of the information, thus yielding a slower rate than would be measured internally within a design office.

Bottleneck indicator (BN): Although the WIP index is an early indicator of potential bottlenecks, it is not sufficient in itself, because designers may view or download information, introducing it into their process and in so doing resetting the WIP index to zero, but then delay use of that information. An additional measure that helps indicate a possible bottleneck is the information development velocity. Where it is low, and has remained low for some time, a bottleneck is likely. Thus taken together, the WIP and DV indices, evaluated either for a set of work packages or for a designer, are a better indicator of a possible bottleneck. The BN indicator is set algorithmically using normalized WIP and DV values (Tribelsky and Sacks, 2010).

Rework: calculated only in retrospective review of design processes, requiring painstaking analysis of the data by the researcher. Its value is expressed as the proportion of information changed to the total information developed (added and changed) between any two steps. In the analysis, negative rework, which results from design errors, erroneous information, or lack of skill or knowledge, is distinguished from positive design iteration, which adds value to the project.

LEAN DESIGN

Lean construction design research has identified a number of prescriptive actions, some of which relate to management of information flows. Formoso *et al.* (1998) suggested value stream analysis to remove non-value-adding tasks, strong focus on acquiring client requirements, stabilization of tasks with high variability, reduction of design cycle times, simplification of the process, increasing the transparency of the process itself to its participants, emphasis on completion of design stages before commencing subsequent steps, and definition of performance measures and monitoring them through the design process. A set of guiding principles for lean design listed by Ballard and Zabelle (2000) include (a) small information batches, that is, sharing even partial information with all of the team, (b) multidisciplinary design teams that also include end-user representatives, (c) design within fixed distinct modules within a building, (d) simultaneous engineering (product) design and construction (process) design, (e) transfer of detailed design of specialized items to fabricators and (f) use of pull flow control. Principles (a) and (f) relate directly to the flow of information.

INFORMATION TECHNOLOGIES

The use of information technology in general, including building information modelling (Eastman *et al.*, 2008), brings numerous benefits to civil engineering projects (Back and Moreau, 2000;Sacks *et al.*, 2003; Nitithamyong and Skibniewski, 2006):

- improvement in the quality of design documents;
- reduction of design cycle times;
- improved budget and cost control;
- better access to information;
- easier monitoring of the designers' activity;
- improved material procurement systems;
- collaboration between designers regardless of their locations;
- better project performance (lower cost, shorter durations, improved quality);
- better management of design changes.

Thomas *et al.* (2004) reported a 4% average reduction in project costs, significantly shortened project durations and overall improved project performance. Stewart and Mohamed (2004) demonstrated strong correlation between construction companies' performance and their implementation of information technology.

More specifically, the use of central server systems on the internet for storing and sharing electronic information in construction projects improves information flow and supports closer collaboration among project teams (Rojas and Songer, 1999; Nitithamyong and Skibniewski, 2006). The use of these tools is commonplace for medium and large projects.

RESEARCH METHOD AND RESULTS

The research sought to characterize the information flows in multidisciplinary design projects according to lean flow measures, and to identify possible relationships between problems in project information flows and the eventual outcomes of those projects. This required examination of a set of comparable projects, recording of their information flows, analysis of the flows according to a set of lean flow measures, independent assessment of the project outcomes and analysis of the results.

Figure 1 provides an overview of the procedure. The first step was to identify an appropriate sample set of projects. Data describing the information flows were recorded through the construction detailing design phase for each project by analysing the archives of their documents and the logs of the document transfers as recorded on the project extranet collaboration website, and by observing design coordination meetings. Data describing the outcomes were collected from post-project interviews with project team managers and designers and from the budget and schedule outcomes. Finally, the two data sets were compared. Each of these stages is described in detail in the following sub-sections.

SELECTING THE SAMPLE PROJECTS

Two principal criteria guided selection of the project sample set. First, the projects should be similar in their context, so that the influence of external parameters (such as owner preferences, procurement methods, timing, etc.) on the design information flows can be factored out by virtue of being the same or very similar. Second, the projects must be current so that the design meetings can be observed and the data describing their flows must be logged reliably, precisely and in detail.

These conditions were fulfilled by selecting a sample composed of the 14 largest sub-projects from a major airport relocation project. The entire project was estimated at $250m and included runways, hangars, service facilities, technical areas, flight infrastructures, water and sewerage infrastructures, operational infrastructures, electrical systems, communication power stations, facilities, fuel lines, administrative areas and living quarters. The 14 sub-projects selected are listed in Table 2 and included civil works, building construction and infrastructure systems sub-projects, and together accounted for $107.2m (43%) of the total. The client had mandated use of an online project extranet, so that all transfers of project documents were logged. The first author's affiliation to the client's project team provided him a unique degree of access to the projects and their design teams, which included open

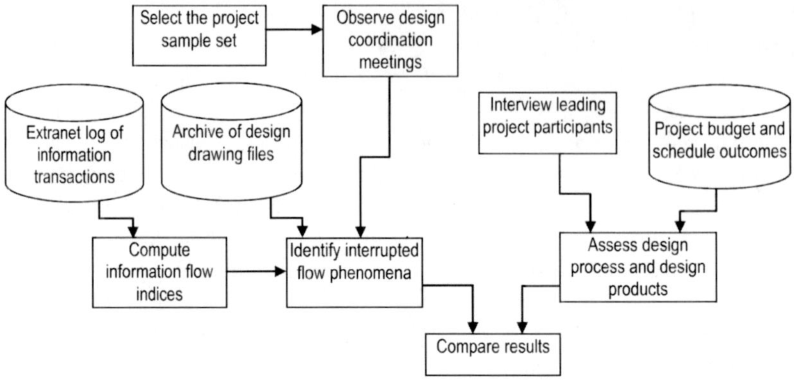

FIGURE 1 The research procedure

access to all project files and communications on the project extranet, open access to all other documents (such as budgets and schedules), unobtrusive participation in project design meetings and interviews with the designers.

Since all the projects were procured by the same client under the same set of constraints and policies, they had very similar profiles. However, each project had a different set of subcontracted design firms (14 on average) and a different subcontracted project manager. All of the designers were selected through closed tender procedures based on qualifications and current workload, not price.

OBSERVATIONS OF DESIGN COORDINATION MEETINGS

The first author participated in 86 design progress meetings for the 14 projects. The meeting durations ranged from 1.5 to 3 hours, were attended by the same 7–10 participants (owner representative, project manager and designers) each time, and were held at approximately 2-week intervals for each project. Because the researcher was an employee of the owner organization, he was able to overcome the difficulties faced by many researchers of communication in construction of gaining free access to meetings and permission to record exchanges (Emmitt and Gorse, 2006). The researcher prepared a detailed protocol of the discussions, examined the documents that were used, and recorded each participant's behaviour in terms of sharing or withholding the design information they held using a pre-prepared form. Wherever classification of an exchange was in doubt, the researcher clarified the issue with the relevant participant at the end of the meeting. These observations enabled collection of complementary information concerning design contents (design programme maturity) and contexts (intensity of design review and geographical dispersion), the dynamics of the participants and their activities (degree of cooperation, leadership style, team motivation) and the general progress of each project. In the context of this article, these were important in assessing the background to the occurrence of information flow bottlenecks, large batch sizes, etc., and enabled confirmation that all recorded design information was indeed exchanged using the project extranet (see below).

EMPIRICAL DATA FROM THE EXTRANET TRANSACTION LOG

Empirical data were collected for each project from the communication and file transfer data log that was compiled automatically by the commercial project management extranet application used to manage the projects. The teams used the application for centralized management of all project information, including CAD files, meeting summaries, presentations, time tables, task assignments and additional information. For each transaction, the log recorded the time at which it was performed, the name of the actor and the file names where relevant. The transaction types were uploading, downloading, updating the version of,

TABLE 2 Profiles of the projects examined

	PROJECT TYPE	DESCRIPTION OF FACILITY	BUILT AREA (m^2)	CONSTRUCTION BUDGET (MILLION $)	NUMBER OF FILE TRANSACTIONS	TIME SPAN OF TRANSACTIONS (DAYS)	NUMBER OF DESIGN FIRMS/ DISCIPLINES IN PROJECT TEAM
A	Residential buildings	A complex of 15 two-storey houses	5,000	4.5	4,870	182	14
B	Runway	A double 3,200 m long runway, 2 × 35 m wide, full structure	230,000	26.5	3,650	545	16
C	Airplane park	Aircraft parking surface	250,000	20.7	13,355	188	15
D	Sewage pumping station	A three-storey building, pumping system, air treatment facilities	320	1.4	2,457	456	12
E	Control tower	An eight-storey control tower above one office floor	800	4.2	6,515	303	16
F	Communication centre	A single-storey building	1,300	3.0	5,627	365	16
G	Infrastructure facilities	Water, waste, electricity and communication networks	NA	22.4	6,879	527	11
H	Office building	A single-storey sheltered office building	1,320	4.2	4,409	226	13
I	Workshops	A complex of three workshop buildings	2,500	2.8	3,232	430	16
J	Office building	A single-storey office building	340	1.1	3,488	487	14
K	Airplane hangar	A hangar for two Boeing 767 airplanes with a two-storey workshop annex	7,300	8.5	4,734	462	16
L	Fuel delivery pipeline	A 2,600 m long fuel pipeline and electrical anti-corrosive system	NA	2.5	4,114	466	11
M	Office building	A single-storey office building with laboratories and stores	1,900	3.9	2,470	282	15
N	Residential building	A three-storey residential building	1,700	1.5	4,248	182	11

viewing, erasing, moving and copying files. The journal logs provided a large sample of documented, detailed and reliable information over extended periods for all 14 projects, with a total of 70,048 transactions recorded and used. Approximately 90% of these were transfers of drawings and schedules, mostly in DWG and some in PDF format. Eight per cent were transfers of technical specification documents and the remaining 2% included meeting summaries, requests for information, client's memoranda and budget directives.

Although use of the extranet system was mandated by the project client, it was important to check that the majority of design information was being exchanged through it in the projects selected. A survey conducted among the designers, reported in Tribelsky and Sacks (2010), showed that this was indeed the case, and this was confirmed in the observations of design progress meetings discussed above.

ANALYSIS OF CHANGES IN ARCHIVED DESIGN DOCUMENTS

In addition, selected drawings from each project were analysed for changes to provide the data necessary evaluate design information development. This was necessary because the transaction records provided no information about the nature of the files' contents. All the versions of every drawing uploaded to the system throughout the life of the projects were available. By comparing a drawing version with its predecessor and by tracking the changes in the drawings using automated review and compare software (Autovue, Cimmetry Systems, 2006), the quantity and the nature of the changes made from version to version were examined and evaluated.

COMPUTE INFORMATION FLOW INDICES

The raw monitoring data for information flows of the 14 projects were used to compute the information flow indices according to the method detailed in the sub-section 'Measuring Design Information Flows' above. Figure 2 shows an example of the batch size index for project A.

IDENTIFY INTERRUPTED FLOW PHENOMENA

The plots of the indices' values were then analysed to determine the frequency and intensity of occurrence of five key waste phenomena (in lean terms), and finally to determine a single relative value for the frequency of the waste phenomena in each project. The results are summarized in Table 3. Projects F (communication centre) and G (roads) had smooth information flows. Projects E (control tower), M (office building) and I (workshops) exhibited all the waste phenomena, albeit with different levels of intensity.

ASSESS DESIGN PROCESS AND DESIGN PRODUCTS

The success or failure of a civil engineering design process can be measured in numerous ways, such as:

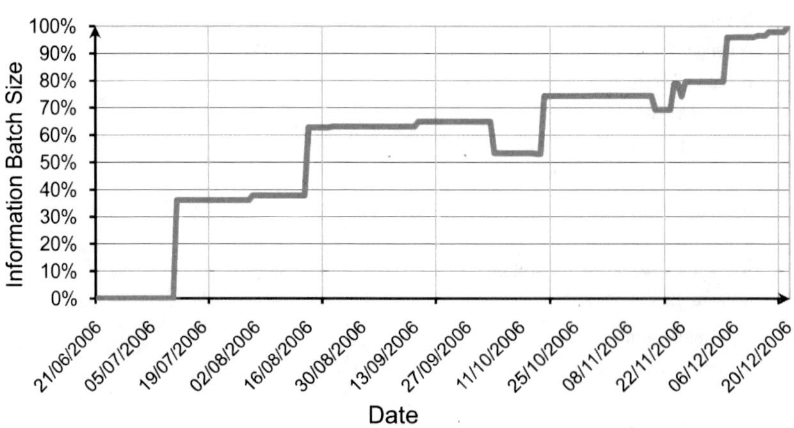

FIGURE 2 Information batch size index monitored over time for project A

TABLE 3 Waste phenomena identified from information flow indices

PHENOMENA OBSERVED PROJECT	BOTTLE-NECKS	REWORK	LARGE BATCH SIZES	HIGH LEVELS OF WIP	FAILURE TO MEET CLIENT NEEDS	FREQUENCY OF WASTE PHENOMENA
A	–	○	•	•	○	6
B	–	•	–	○	○	4
C	–	•	○	○	–	4
D	•	–	○	○	–	4
E	○	•	○	○	○	6
F	–	–	◙	◙	–	0
G	–	–	◙	◙	–	0
H	–	–	○	○	–	2
I	•	•	•	○	•	9
J	○	–	•	○	○	5
K	•	○	–	–	○	4
L	•	•	–	○	–	5
M	○	•	○	○	○	6
N	–	–	○	•	–	3

Notes:

•Strong (2): the phenomenon appears consistently throughout the project duration.

○Weak (1): the phenomenon appears clearly, but with low intensity.

–None (0): the phenomenon was not observed in the data.

◙Opposite: the inverse phenomenon occurred (i.e. small batch sizes, low WIP).

- measuring the quality of the design 'products', in terms of constructability, design errors and inconsistencies in the models, drawings or other documents;
- assessing the participants' satisfaction with the process (amount of rework, meeting design schedule and estimated cost targets, extent of teamwork and collaboration);
- measuring the degree to which the target project budget and design durations are met once design is complete;
- measuring the degree to which construction budget, schedule, quality and safety targets are met;
- assessing the constructed facility's ability to fulfil the functional requirements defined in the brief or programme; and
- assessing the long-term satisfaction of the projects' users with the constructed facility.

The latter three are of course dependent not only on the design process, but also on multiple other factors, such as project management, contractor performance and external constraints. Although all are useful, they cannot be aggregated into any single measure of project success. The time span of the research precluded use of the latter three measures, and so the approach adopted was to use all of the first three and to compare each of them with the flow measures computed in the previous steps.

The design documents were assessed using three methods:

1 Analysis of the design documents produced using a method developed by Chang and Ibbs (1999). According to this method, the evaluator assesses the drawings using a weighted scale of 11 attributes, classified into groups for constructability, usability and accuracy. In this research, each attribute of the set of drawings in each project was evaluated on a four-point scale. The results are shown in Table 4.
2 Analysis of the results of independent design reviews performed for each project by client-appointed experts as part of the quality assurance system. Each expert reviewed the

TABLE 4 Assessment of the effectiveness of design documents using Chang and Ibbs' (1999) method

PROJECT	CONSTRUCTABILITY					USABILITY			ACCURACY			
	ADAPTABILITY	MAINTAINABILITY	CONSTRUCTION KNOWLEDGE	CONST. METHOD AND TIME	FORMAT	CONFORMANCE	CLARITY	COMPLETENESS	IMPACTS OF ERROR	NO ERRORS AND NO OMISSIONS	CONSISTENCY	WEIGHTED AND NORMALIZED EFFECTIVENESS
A	–	○	○	–	○	–	○	○	•	○	•	37
B	•	○	–	•	–	•	○	–	–	–	○	51
C	–	○	•	○	○	○	○	•	–	○	○	63
D	–	•	•	–	•	–	•	•	○	•	•	86
E	–	•	•	○	•	○	•	○	–	○	○	74
F	○	•	•	–	•	○	○	•	○	•	•	91
G	•	•	○	•	○	○	○	○	○	○	○	80
H	•	•	○	○	•	○	○	•	–	○	○	80
I	○	○	○	–	•	○	○	○	•	–	○	63
J	○	–	○	○	○	•	•	•	•	•	•	63
K	–	•	•	○	•	○	•	•	○	○	○	86
L	–	–	•	○	•	○	•	•	–	○	–	63
M	○	○	○	○	•	○	○	○	○	○	–	63
N	○	○	○	–	•	○	•	•	–	•	•	80

Legend:

• Strong.

○ Medium.

– Weak or none.

designs for the discipline of their expertise across all 14 projects. The number of critical issues identified for each project, for the two main design disciplines relevant for each project, were counted and used to calculate normalized design quality review scores for each project. The results are summarized in Table 5, together with the design product assessments from the following step.

3 Thirteen interviews were conducted with client representatives and with project managers in which they were asked to rate their degree of satisfaction with the design drawing sets. The scale ranged from 'very dissatisfied' (1) to 'fully satisfied' (5). The client representatives each dealt with some 40 projects within the overall airport project, and the project managers with some 10 each, so that they were able to evaluate the projects they were involved in within a broad context (the interviewees included the project managers for all 14 projects and the owner representatives for nine of them, yielding a total of 23 responses from 13 interviews). The researcher who participated in the design review meetings and examined the documents provided a third opinion for each project (see Table 6). These values are termed 'subjective' in the discussion below.

The design process itself was also assessed using subjective and objective measures. In the interviews described above, the client representatives and the project managers assessed their satisfaction with the design process using the same five-point scale, providing subjective measures. Objective measures were obtained by computing the ratio of the final detailed design estimate of the construction cost to the initial project budget, and the ratio of the actual to the planned design phase duration, for all of the projects. Before calculating the ratios, any additional costs and design duration extensions that resulted from client-initiated changes after the initial estimates were made were deducted from the final values. The results of these steps are listed in Table 6.

ANALYSIS AND DISCUSSION

The research question posed the hypothesis of correlation between the nature of the design information flows in a project, assessed according to

TABLE 5 Peer review and participant assessments of the design drawings

PROJECT	INDEPENDENT PEER DESIGN REVIEWS				PARTICIPANT'S ASSESSMENTS		
	PRIMARY DISCIPLINE DESIGN REVIEW[a]		SECONDARY DISCIPLINE DESIGN REVIEW		PROJECT MANAGERS	CLIENT REPRESENTATIVES	RESEARCHER
	CRITICAL ISSUES/ TOTAL ISSUES	SCORE	CRITICAL ISSUES/ TOTAL ISSUES	SCORE			
A	12/24	7.1	3/9	9.2	3	–	3
B	8/27	5.7	–	–	3	–	2
C	14/25	9.9	32/48	17.7	4	–	3
D	2/18	7.8	–	–	2	4	2
E	12/25	7.1	18/54	9.9	3	4	3
F	14/63	8.3	1/4	10.2	3	5	3
G	1/4	9.3	–	–	1	2.5	2
H	18/23	10.7	32/54	17.7	3	2	3
I	3/15	17.5	18/60	13.9	2	5	2
J	4/10	5.9	3/10	9.1	4	5	3
K	5/12	5.8	2/10	10.2	4	5	3
L	–	–	–	–	5	4	4
M	0/1	0	3/5	1.7	3	4	2
N	8/22	4.7	1/2	0.6	4	–	4

[a]Ratio of critical issues to total issues.

TABLE 6 Design process review results, ratio of final design cost estimate to planned construction cost and ratio of actual design to duration to planned duration

PROJECT	SATISFACTION WITH THE DESIGN PROCESS			BUDGET OVERRUN (%)	SCHEDULE OVERRUN (%)
	PROJECT MANAGERS	CLIENT REPRESENTATIVES	RESEARCHER		
A	2	–	3	−2.3	−47.6
B	4	–	3	−34.3	124.1
C	4	–	3	5.8	−5.4
D	3	2	1	13.7	−21.5
E	4	4	1	−2.0	22.0
F	2	2	3	1.4	−18.6
G	2	–	2	−20.3	2.0
H	3	4	1	−16.4	0.0
I	4	4	1	−29.1	10.4
J	4	5	1	−10.0	117.7
K	4	4	2	−10.7	127.0
L	4	3	3	−6.9	181.6
M	4	3	3	−3.0	−23.2
N	3	–	4	2.1	37.3
Average	3.3	3.2	3.1	−3.1	36.1
Median	4	3.5	3.5	−2.7	6.2

the lean production conceptualization of flow, and the design process outcomes. The data defining the occurrence of interruptions to information flow in the projects, summarized in Table 3, describe the nature of the information flows. Evaluation of the hypothesis requires comparison of these data with the set of design process outcome measures detailed in Tables 4–6. Each of the following sub-sections presents a comparison between the data in Table 3 and one of the outcome assessments.

PROCESS FLOW AND DESIGN DOCUMENT QUALITY

In Figure 3, the frequency of flow interruption phenomena data from Table 3 is plotted against the design document measure results calculated using Chang and Ibbs' method and listed in Table 4. The absolute value of the correlation coefficient of the two sets of numbers is 0.61. This indicates that where more information flow waste phenomena were observed (where information flow was impeded), the design documents produced by the team tended to be of lower quality than for the projects where information flowed freely. The trend line computed for the data clearly shows the relationship between the two. This finding echoes the relationship between end product quality and variable or unstable production flow that is symptomatic of industrial production systems (Hopp and Spearman, 1996). On the other hand, no clear

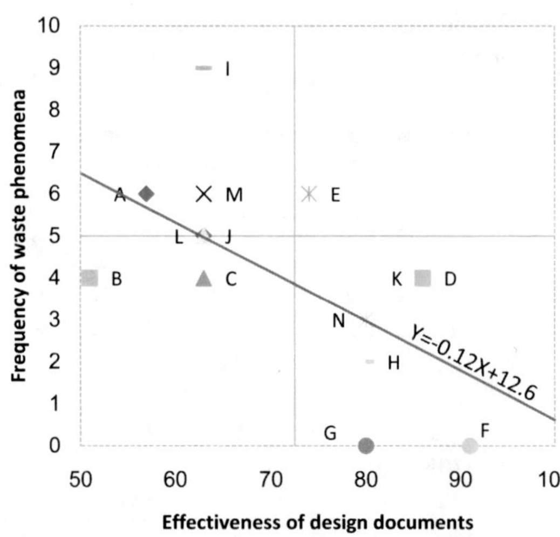

FIGURE 3 Relationship between Chang and Ibbs' measure of the effectiveness of design documents and the frequency of occurrence of information flow waste phenomena

correlation was found between the process flow interruption phenomena and the subjective reviews of design product satisfaction for any of the three evaluations.

PROCESS FLOW AND DESIGN PROCESS MEASURES

The charts in Figure 4 plot the data for the information flow waste phenomena against the degree to which each project met its budget and schedule targets. As the quality of the information flow worsens from project to project, the spread of the deviations for both budget and schedule grows. In other words, projects with smooth information flows appear more stable than those with interrupted or inefficient information flows. The projects with poorer information flows were less predictable, in terms of both their duration and the degree to which they met their initial construction budget targets. This observation conforms to the predictions of lean production flow theory, where the correlation between instability in flows within a process and

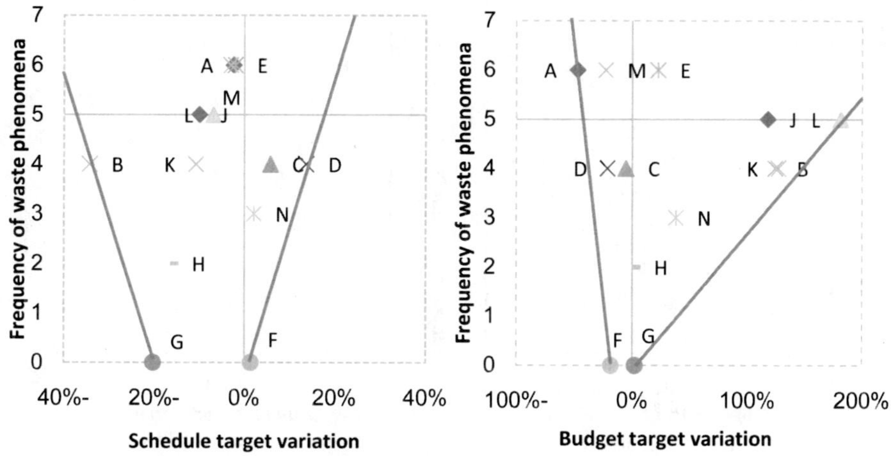

FIGURE 4 Relationships between variation in meeting schedule and budget targets and the frequency of information flow waste phenomena

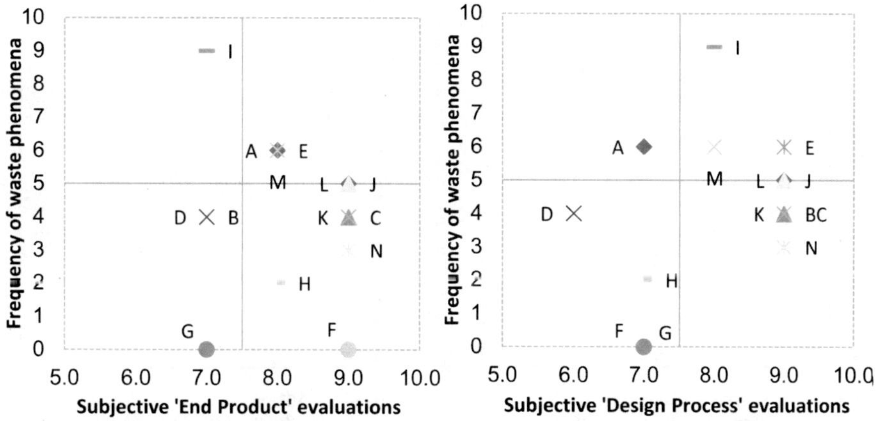

FIGURE 5 Relationships between subjective evaluations of 'End Product' and 'Design Process' quality and the frequency of information flow waste phenomena

lack of predictability in the results of the process is well understood (Koskela, 2000).

QUALITY AND THE FREQUENCY OF INFORMATION FLOW WASTE PHENOMENA

In this case too, no clear correlation was found between the process flow interruption phenomena and the subjective reviews of design process on the part of the client and the project managers (see Figure 5). However, a small positive correlation was found between the phenomena and the researcher's review. The distinct lack of correlation between the project managers' subjective assessments of the design processes they managed and the actual nature of their information flows suggests that the managers were either unaware of the negative impact of poor information flows on their projects, or unwilling to admit the effect.

In this context, it is interesting to consider the relationship between the subjective reviews of the

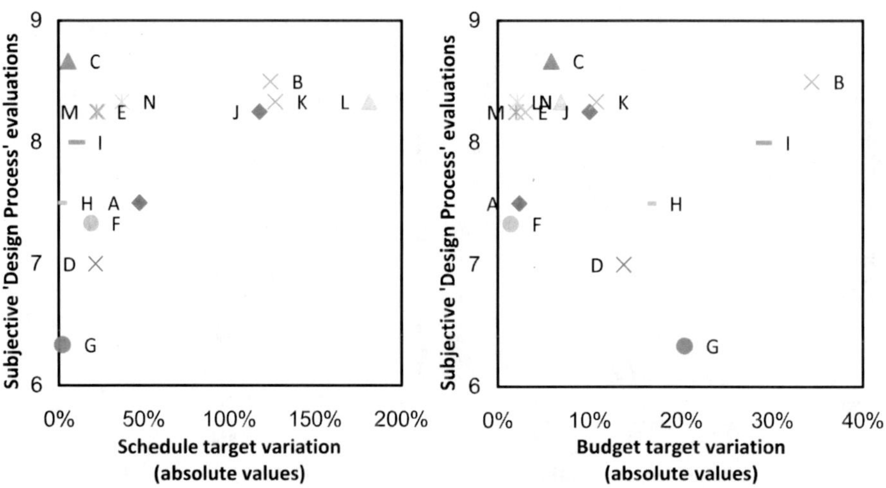

FIGURE 6 Relationships between subjective 'Design Process' evaluations and schedule and budget variations

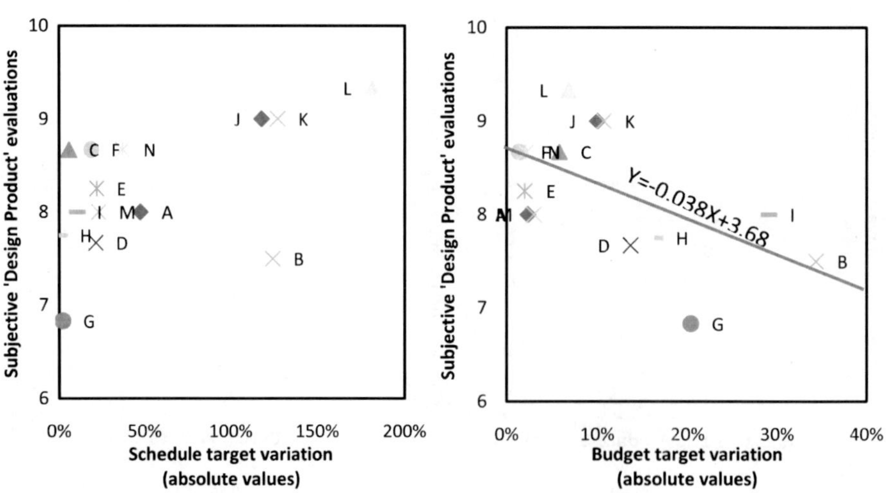

FIGURE 7 Relationships between subjective 'Design Product' evaluations and schedule and budget variations

design processes and the degree of success in meeting budget and schedule targets. The relevant data are plotted in Figure 6.

It appears that the subjective reviews of the design process itself bear little or no relationship to the apparently objective measures of schedule and budget deviation. Thus little can be learned from them concerning evaluation of the information flows, apart from the observation that positive reviews of a design process on the part of either project managers or clients do not necessarily indicate that the projects' information flows are healthy. Rather, the independent information flow indices (Table 2), computed on the basis of the log of file transactions by the design team, are better indicators of smooth or interrupted information flows.

Similarly, no correlation was found between the subjective perceptions of success in managing the design process and meeting the planned design process duration (Figure 6). There is, however, clear correlation between the subjective perceptions of success in achieving the required design product goals and meeting budget targets (the absolute correlation coefficient is 0.56, and the trend line is shown in Figure 7).

CONCLUSIONS

This study sought evidence of a possible correlation between the characteristics of information flow between design team members and the degree of success achieved in the detailed design phase of civil engineering projects. A unique sample of 14 independent projects (including housing, office buildings, runways, hangars and infrastructure), all part of a major airport relocation project, was studied. All the projects had the same client and design procurement methods, and all used the same information technologies, but they had different project managers and design team members. The primary differences between them were therefore limited to their management style, group dynamics and the behaviours of design team members.

The information flows within each design team were characterized in terms of measures drawn from the concepts of lean thinking, such as flow bottlenecks, rework, large batches and cycle time, which were monitored and computed analytically using a set of seven information flow indices. The degree of success was measured in terms of the effectiveness of the design documents that were produced, the projects' success or failure in meeting targets of budget, schedule and client expectations, and supplemented by subjective assessments collected in interviews with key personnel from all the projects.

The results show a direct relationship between the objective measures of information flow and the measures of the effectiveness of design documents. Stable design information flows, with small batches, frequent transactions and lower occurrence of flow interruption phenomena lead to higher quality design documents. Similarly, the degree to which projects vary in terms of meeting budget and schedule targets appears to correlate with the frequency of occurrence of phenomena that indicate disruption of smooth information flows. This means that the probability for projects that have unstable flows to exceed their planned design durations or budgets is higher than for those with stable flows. Unstable flows are not necessarily the root cause of the unpredictability of outcomes, but they are at least a predictor for it.

However, correlation between subjective impressions of a project's success on the part of its participants, including the client, and the quality of information flows could not be proven. This was true both for their assessments of the final product and of the design process itself (Figure 5). Interestingly, there was similarly no clear correlation between the design teams' evaluations of their own successes and the degree to which they succeeded in meeting budget or schedule targets.

The positive correlation between the occurrence of information flow bottlenecks, infrequent transactions involving large batch sizes, large quantities of WIP and significant rework, on the one hand, and low effectiveness of design documents on the other hand, supports the validity of the lean production flow conceptualization of the information flows in detailed design of civil engineering projects. The correlation between these lean measures of flow and the predictability with which teams succeed or fail to achieve their budget or schedule targets provides further support for this conclusion.

The operational conclusion is that design managers can improve processes by paying attention to the lean characteristics of the information flows in their projects (encouraging small batch sizes, maintaining low quantities of WIP, short design review and response cycle times, identifying and releasing bottlenecks). Although recognizing that a design team's success in producing effective design documents while meeting budget and schedule constraints is dependent on numerous factors, some of which are beyond the team's control, it appears that team members' behaviour in sharing information is an important determinant of success. Among the teams studied those that exploited the collaboration website to implement frequent and smooth information flows produced better design products.

REFERENCES

Alexander, C.W., 1964, *Notes on the Synthesis of Form*, Harvard University Press, Cambridge.

Ashford, J.L., 1989, *The Management of Quality in Construction*, E & FN Spon, London.

Back, W.E. and Moreau, K.A., 2000, 'Cost and schedule impacts of information management on EPC process', *ASCE Journal of Management in Engineering* **16**(2), 59–70.

Baldwin, A.N., Austin, S.A., Hassan, T.M. and Thorpe, A., 1999, 'Modelling information flow during the conceptual and schematic stages of building design', *Construction Management and Economics* **17**(2), 155–167.

Ballard, G., 1998, 'Positive vs. negative iteration in design', *Sixth Annual Conference of the International Group for Lean Construction (IGLC-6), Guaruja, Brazil*, 12.

Ballard, G., 2000, 'The last planner™ system of production control', PhD dissertation, The University of Birmingham, Birmingham, UK.

Ballard, G. and Zabelle, T., 2000, *Lean Design: Process, Tools, & Techniques*, Lean Construction Institute, La Jolla, CA.

Chang, A. and Ibbs, C.W., 1999, 'Designing levels for A/E consultant performance measures', *Project Management Journal* **30**(4), 42–55.

Eastman, C.M., Teicholz, P., Sacks, R. and Liston, K., 2008, *BIM Handbook: A Guide to Building Information Modeling for Owners, Managers, Architects, Engineers, Contractors, and Fabricators*, John Wiley and Sons, Hoboken, NJ.

Eckert, C., Clarkson, J. and Stacey, M., 2001, 'Information flow in engineering companies: problems and their causes', in: S. Culley (ed), *International Conference on Engineering Design ICED 01*, John Wiley and Sons, Glasgow, 43–50.

Emmitt, S., 2007, *Design Management for Architects*, Blackwell Publishers, Malden, MA.

Emmitt, S. and Gorse, C., 2006, *Communication in Construction Teams*, Spon Press – Taylor and Francis, Abingdon.

Ford, D.N. and Sterman, J.D., 2003, 'The Liar's Club: concealing rework in concurrent development', *Concurrent Engineering: Research and Applications* **11**(3), 211–219.

Formoso, C., Tzortzopoulos, P., Jobin, M. and Liedtke, R., 1998, 'Developing a protocol for managing the design process in the building industry', *Proceedingds of the 6th Annual Conference of International Group for Lean Construction, Berkeley, CA*, International Group for Lean Construction.

Freire, J. and Alarcon, L.F., 2002, 'Achieving lean design process: improvement methodology', *Journal of Construction Engineering and Management* **128**(3), 248–256.

Fyall, M., 2002, *When Project Information Flow Becomes Turbulent: Toward an Organizational Reynolds Number, 138*, CIFE – Center for Integrated Facility Engineering, Stanford, CA.

Goldratt, E.M., 1997, *Critical Chain*, North River Press, Great Barrington, MA.

Gray, C. and Hughes, W., 2001, *Building Design Management*, Butterworth-Heinemann.

Hopp, W.J. and Spearman, M.L., 1996, *Factory Physics*, IRWIN, Chicago, IL.

Huovila, P., Koskela, L. and Lautanala, M., 1997, 'Fast or concurrent: the art of getting construction improved', in: L.F. Alarcón (ed), *Lean Construction*, Balkema, Rotterdam, The Netherlands, 143–159.

Koskela, L., 2000, *An Exploration Towards a Production Theory and its Application to Construction*, D. Tech, Helsinki University of Technology, Espoo.

Koskela, L.J., Huovila, P. and Leinonen, J., 2002, 'Design management in building construction: from theory to practice', *Journal of Construction Research* **3**(1), 1–16.

Krovi, R., Chandra, A. and Rajagopalan, B., 2003, 'Information flow parameters for managing organizational processes', *Communications of the ACM* **46**(2), 77–82.

Love, P.E.D., Holt, G.D., Shen, L.Y., Li, H. and Irani, Z., 2002, 'Using systems dynamics to better understand change and rework in construction project management systems', *International Journal of Project Management* **20**(6), 425–436.

Love, P.E.D., Mandal, P., Smith, J. and Li, H., 2000, 'Modeling the dynamics of design error induced rework in construction projects', *Construction Management and Economics* **18**(5), 567–574.

Manavazhi, M.R. and Xunzhi, Z., 2001, 'Productivity oriented analysis of design revisions', *Construction Management and Economics* **19**, 379–391.

Moreau, K.A. and Back, W.E., 2000, 'Improving the design process with information management', *Automation in Construction* **10**, 127–140.

Morgan, J. and Liker, J.K., 2006, *The Toyota Product Development System: Integrating People, Process and Technology*, Productivity Press, New York.

Nitithamyong, P. and Skibniewski, M.J., 2006, 'Success/failure factors and performance measures of web-based construction project management systems: professionals' viewpoint', *Journal of Construction Engineering and Management* **132**(1), 80–87.

Ostergaard, K.J. and Summers, J.D., 2007, 'Resistance based modeling of collaborative design', *Concurrent Engineering* **15**(1), 21–32.

Prasad, B., Morenc, R.S. and Rangan, R.M., 1993, 'Information management for concurrent engineering: research issues', *Concurrent Engineering* **1**(1), 3–20.

Rojas, E.M. and Songer, A.D., 1999, 'Web-centric systems: a new paradigm for collaborative engineering', *Journal of Management in Engineering* **15**(1), 39–45.

Sacks, R., Eastman, C.M. and Lee, G., 2003, 'Process improvements in precast concrete construction using top-down parametric 3-D computer-modeling', *Journal of the Precast/Prestressed Concrete Institute* **48**(3), 46–55.

Steward, D.V., 1981, 'The design structure system: a method for managing the design of complex systems', *IEEE Transactions on Engineering Management* **28**, 71–74.

Stewart, R.A. and Mohamed, S., 2004, 'Evaluating web-based project information management in construction: capturing the long-term value creation process', *Automation in Construction* **13**(4), 469–479.

Thomas, S.R., Lee, S.-H., Spencer, J.D., Tucker, R.L. and Chapman, R.E., 2004, 'Impacts of design/information technology on project outcomes', *Journal of Construction Engineering and Management* **130**(4), 586–597.

Tilley, P.A., 2005, 'Lean design management – a new paradigm for managing the design and documentation process to improve quality?', in: R. Kenley (ed), *13th Conference of the International Group for Lean Construction, University of New South Wales, Sydney, Australia*, 283–295.

Tribelsky, E. and Sacks, R., 2007, 'Measures of information flow for lean design in civil engineering', in: W. Hughes (ed), *Proceedings of CME25 Construction Management and Economics: Past, Present and Future*, Taylor and Francis, University of Reading, London, UK.

Tribelsky, E. and Sacks, R., 2010, 'Measuring information flow in the detailed design of construction projects', *Research in Engineering Design* **21**(3), 189–206.

Management of Waste in the Building Design Process: The Ghanaian Consultants' Perspective

E. Zoya Kpamma[1] and T. Adjei-Kumi[2]

[1]Department of Building Technology, Sunyani Polytechnic, Sunyani, Ghana
[2]Department of Building Technology, Kwame Nkrumah University of Science and Technology, Kumasi, Ghana

Abstract

Waste associated with building design is one of the causes of the high cost and slow progress of construction projects in Ghana. Proper management of waste at the design stage is therefore a fundamental step towards achieving speedy delivery of building projects at minimum cost. This paper seeks to examine the level of awareness of sources of waste and the application of waste reduction tools in the building design process by Ghanaian consultancy firms. Data on the design activities of the firms were collected by distributing questionnaires to consultants, while contractors and clients were interviewed. Involvement in various building design processes was also used to obtain data for the study. Results revealed a generally low recognition of sources of waste in the building design process and little awareness of waste reduction tools such as design structure matrix, batch size reduction and set-based design. It was also found that inadequate familiarity of the firms with lean thinking was among a number of limitations to the application of waste reduction tools in Ghana. This suggests that there is a need to improve the awareness of lean design management in both education and practice in Ghana.

INTRODUCTION

One of the banes of the construction industry in Ghana is the inability of firms in the industry to deliver products and services on time and within lower cost limits. Construction project delivery in Ghana is seen to be associated with high cost and time overruns as well as poor quality (Westring, 1997; World Bank, 1996, 2003). A survey conducted by Nicco-Annan (2006) on the construction of some office buildings in Accra reveals cost overruns of between 60 and 180%, not taking inflation into account. Time overruns of between 12 and 24 months were also observed in a survey conducted on the construction of some office buildings in Accra (Nicco-Annan, 2006).

The growing need for speedy delivery of value, associated with less waste, in the construction project delivery process makes it imperative for firms in the construction industry in Ghana to adopt measures towards reducing waste and maximizing value for their clients in the project delivery process. A great deal of attention is given to design management in the architecture–engineering–construction (AEC) sector due to its strong impact on the entire project (Chua and Tyagi, 2001). Several investigations have indicated that a large percentage of defects in buildings arise from decisions or actions at the design phase (Cornick, 1991). It has also been established that design has a lot of impact on the level of efficiency during the production stage of construction projects (Ferguson, 1989).

According to Undurraga (1996), about 20–25% of total construction time is wasted as a result of design deficiencies. Around 78% of quality problems of AEC projects are design related (Koskela, 1992). In relation to cost overruns, design-related causes constitute the major category (Josephson and Hammarlund, 1996). In the case of Ghana, Westring (1997) attributes delays in construction projects, among others, to delays in preparation of drawings and technical specifications.

One of the problems that design management practice has to address is the waste associated with the design process. Seemingly intangible and subtle, waste in design arises out of delays, waiting, design errors, overprocessing and negative iteration (Ballard, 2000). The waste arising out of these sources in the design process can have far-reaching implications by undermining efforts towards delivering valuable construction products on time to clients. This article will review some waste-in-design management practices and discuss the results of a survey conducted on the operations of some consultancy firms in Ghana to assess the extent to which they are familiar with waste management practices in the building design process in line with lean thinking.

DESIGN MANAGEMENT AND WASTE

DESIGN MANAGEMENT PRACTICE

Melhado (1994) defines building design as 'an activity or service that is an integral part of the building construction process, dealing with development, organization, documentation and transmission of information on the specified physical and technological characteristics that must be considered in the construction phase of a particular project'. To transmit information on the characteristics of a building, in line with this view of building design, requires design to be considered as much more than drawing and description (Melhado, 1998). Design must be considered as a multifunctional process involving not only architectural and engineering tasks but also ranging from marketing and cost analysis to technological choices and production process specifications (Melhado, 1995). Searching for a new way of considering the difficulties mentioned above, at the same time including the necessary changes to achieve efficiency improvement objectives in design, leads us to the need for design management.

The management of design and engineering is generally seen to be problematic in AEC projects (Ballard and Koskela, 1998). The main problems in design management have generally been identified as poor communication, lack of adequate documentation, deficient or missing input information, unbalanced resource allocation, lack of coordination between disciplines and erratic decisions (Cornick, 1991; Austin *et al.*, 1994; Huovila *et al.*, 1997). According to Coles (1990), design problems are significantly a result of poor briefing and communication, insufficiency in the technical knowledge of designers and a lack of confidence in pre-planning for design work. Common effects of this, which could result in waste arising from delays, include slow approvals from clients, late appointments of consultants and inadequate time to complete design documents carefully (Ballard and Koskela, 1998). The most common causes of deviations when designing are deficient planning and/or resource allocation, deficient or missing input information, and changes (Sverlinger, 1996).

MODELS OF DESIGN MANAGEMENT

A number of design management models have been developed (Chua and Tyagi, 2001). One of the benefits of formulating design models is the possibility of effectively planning and controlling the process (Tzortzopoulos and Formoso, 1999). Some of the popular models are Pugh's 'total design' model, the VDI model of engineering design and the Pahl and Beitz design model (Austin *et al.*, 1999). Another commonly used building design model that models the tasks to be performed by various design players during each stage, but does not however model the relationships between the tasks, is the RIBA Plan of Work for Design Team Operation (1973) (Austin *et al.*, 1999).

A variety of modelling techniques have also been employed to model information flow in design (Chua and Tyagi, 2001). They include data flow diagrams, IDEF techniques (including IDEF0), entity-relationship diagrams, object-oriented modelling techniques, Petri nets and a dependency matrix (AdePT) (Austin *et al.*, 2000). Design collaboration frameworks such as design agent-based collaboration (Khedro *et al.*, 1994) and collaborative construction agents (Jones and Riley, 1995) have also been developed.

In Ghana, the design process largely follows the RIBA Plan of Work and is one of the preliminary stages in the entire building construction project delivery process. Traditionally, the design process is preceded by a stage of the client's briefing (or inception) and is broadly made up of the conceptual design stage and the detail design stage. It is at the briefing stage that the intentions and requirements of the client system are established to inform the decisions to be taken at the conceptual and detail design stage (Emmitt, 2002). The process in AEC projects is conventionally led by the architect who transforms the client's abstract intentions into a preliminary layout and the layout used by engineering design specialists to prepare their designs (Chua and Tyagi, 2001).

LINEAR AND ITERATIVE DESIGN PROCESS

According to Emmitt (2002), the conceptual design stage involves an expression of the idea underlying the designer's vision and helps to channel the multitude of design decisions that follow. Concepts, he further explains, embody the architectural and cultural symbolism, and may be expressed in a few sets of lines on a piece of paper that form a sketch, or in a diagrammatic form through to quite elaborate drawings.

The detail design phase is the period during which technology is applied to abstract ideas and concepts to realize the design vision. Practically, the detail design stage is a complicated process made up of the coordination of diffuse information, the management of different consultants and the making of several decisions that will affect the appearance, durability and cost of the product (Emmitt, 2002). During the detail design stage, each engineering discipline works in isolation to prepare its designs, while the design specialists collaborate with each other to prevent any potential conflict or to confirm their designs with one another (Chua and Tyagi, 2001). The confirmation of designs is done after the drawings have been prepared by the design specialists, by which time a number of resources have gone into preparing the drawings and any change needed in them implies expensive rework (Chua and Tyagi, 2001).

Going by the RIBA Plan of Work leads us to a linear model, as shown in Figure 1. Here the design process is adversarial to the principles of lean design (where integration of processes and involvement of participants is key) since the separation between conceptual design, detail design and site operations is definite. The iterative nature of design is also ignored in the model. The traditional design process is plagued by long rework loops, and transparency is at the stage of design whereby a number of resources have been exhausted up to that stage (Chua and Tyagi, 2001). Collaboration is mainly passive and only occasioned by a potential conflict predicted by a design specialist or a conflict discovered during collaboration at the drawings level (Chua and Tyagi, 2001). The traditional design gives priority to task management and puts value management into secondary position (Ballard and Koskela, 1998). All the foregoing issues about the traditional design process motivate the application of lean principles to the design process in building design (Chua and Tyagi, 2001).

The implication of the linear project delivery model, as observed by Emmitt (2002), is that the designer (architect) deals with the creative stage; the technologist (detailer) handles the more technical aspects and contract documentation, while the builder assembles the components specified by the detailer to produce the finished product. This model clearly does not include the manufacturers and suppliers of products at the design stage and could lead to waste generation resulting from delays due to uncertainties during the design and construction of building products.

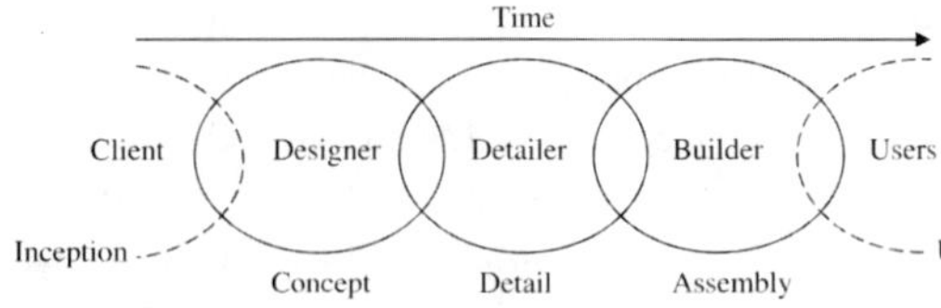

FIGURE 1 The linear project delivery model
Source: Emmit (2002).

According to Ballard (2000), design, by its nature, is an iterative and generative process and that iteration is essential for generating value in the design process (Figure 2). Building design is a difficult process to manage and involves thousands of decisions, probably over a period of years, with several interdependencies, in a highly uncertain environment (Formoso *et al.*, 1998). It has, however, been observed that not all iteration generates value, since iteration that can be eliminated without loss of value is classified as waste (Ballard, 2000).

WASTE IN THE DESIGN PROCESS

According to Ohno (1978), sources of waste in production include overproduction, waiting, transportation, processing, inventory, movement and making defective products. More sources of waste have, however, been suggested by Koskela (2004) and Macomber and Howell (2004). Whereas Koskela (2004) identifies 'make do' as one other source of waste, Macomber and Howell (2004) suggest 'not listening' as another source of waste. According to Koskela (1992), there are four kinds of activities involved in the design process: conversion, waiting, moving and inspection. Only conversion is regarded as a value-adding activity, but waiting, moving and inspection are non-value-adding activities and should be eliminated (Koskela, 1992). Part of the conversion activities are not value adding, since they cause rework due to errors, omissions and uncertainty (Huovila *et al.*, 1997). Design, viewed as an iterative and generative process, has the waste associated with it being explained by Koskela and Huovila (1997) as that which is unnecessary for task completion and value generation. Thus, iteration in design is wasteful and negative when it can be eliminated without loss of value or causing failure to complete the project (Ballard, 2000). Informal surveys of design teams have revealed estimates as high as 50% of design time spent on needless (negative) iteration (Ballard, 2000).

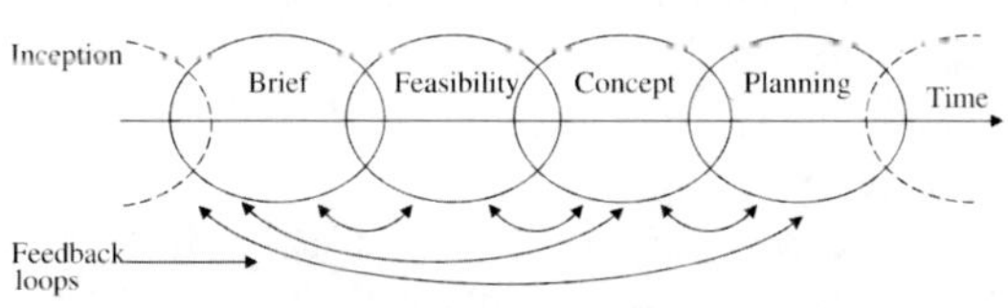

FIGURE 2 The iterative process of design
***Source:* Emmitt (2002).**

Apart from negative iteration, there are other forms of waste associated with the design process. One of them is design errors (Ballard, 2000). According to Reinertsen (1997), design outputs are classified as defective when they fail because something previously known was forgotten or neglected. Design outputs can, however, be failures but not errors if they fail because of a lack of knowledge not previously possessed (Ballard, 2000). Waiting, overprocessing and overproduction are some other forms of waste in design.

REDUCING WASTE IN THE DESIGN PROCESS

Three methods have been proposed for eliminating waste in the design process based on the view of design as a flow (Huovila *et al.*, 1997). The first method is reduction of uncertainty, being one of the main origins of rework particularly in the early stage of design. This method can be implemented by stepping up the effort in spelling out the project boundaries and the requirements of clients. The second method involves reducing waiting time by adequately breaking down the design tasks so that they can be properly planned, and also allowing for the transfer of information in smaller batches. The last method has to do with reduction in effort needed for information transfer. This can be done through teamwork and by rearranging the design tasks.

Negative iteration is an important source of waste in design, but can be minimized by the application of techniques such as team problem solving, design structure matrix (DSM), batch size reduction, least commitment and set-based design (Ballard, 2000). These techniques can be categorized as restructuring the design process, reorganizing the design process, managing the design process differently than traditionally and lastly overdesigning, when there is no better solution (Table 1).

LEAN PROJECT DELIVERY

Lean project delivery builds cooperation in the context of a single integrated team involving the owner,

TABLE 1 Strategies for reducing negative iteration

Restructure the design process
Use DSM to resequence
Use pull scheduling to reduce batch sizes and achieve greater concurrency
Reorganize the design process
Make cross-functional teams the organizational unit
Use team problem solving (call a meeting)
Share ranges of acceptable solutions
Change how the design process is managed
Pursue a least-commitment strategy
Defer this decision (defer commitment)
Practice set-based design
Use the last planner system of production control
Overdesign (design redundancy) when all else fails

Source: Ballard (2000).

architect, constructor and other critical players as equals in the pursuit of a shared goal (Mossman *et al.*, 2010). The lean project delivery system (LPDS) has been designed to support a new and better way of designing and building capital facilities. This model recognizes the design and construction process as both linear and iterative, thus supporting a parallel rather than a sequential approach to design and construction. The various waste reduction strategies, such as DSM, are applied in the lean project delivery process at the lean design phase to ensure minimization of waste in the design process.

RESEARCH METHOD

The research involved representative samples of selected Ghanaian consultancy and construction firms located in various parts of Ghana. This was an opportunity to critically explore and examine the steps in the pre-contract activities of Ghanaian consultancy and construction firms. These stages include site survey, design brief formulation, sketch and detail design, obtaining permits, preparation of bills of quantities, tender documentation and contract documentation.

The LPDS model was used as the basis for assessing the extent to which the various steps of the pre-contract activities (especially the design process) of Ghanaian consultancy firms fall in line with the concept of lean thinking towards waste reduction. The steps were examined and assessed within the broad framework of lean project definition, lean design, lean supply, lean assembly and use. The level of familiarity and application of the various waste-in-design reduction techniques, such as DSM, was also assessed.

TARGET POPULATION

The core target population for data collection using questionnaires consisted of consultants (architects, engineers, quantity surveyors, etc.). The focus on consultants in the distribution of the questionnaire was due to the key role that consultants play in design activities. Contractors (building and civil) and clients (public and private) were also interviewed as a follow-up, specifically to confirm the consultants' responses on the level of participation of clients and contractors in the design process.

PROCEDURE FOR DATA COLLECTION

A sample of 25 firms providing design and documentation services at the pre-contract stage of construction projects across Ghana was selected. Three professionals (architect, quantity surveyor and engineer) in each firm were identified and given the questionnaire. In the case of firms that did not have all the professionals in-house, their associate professionals were contacted and given the questionnaire. Out of a total of 75 questionnaires given out, 68 were returned completed. A personal observation of some activities of some of the selected firms also provided an opportunity to obtain additional information for the study. In addition to the questionnaires that were given out to the consultants, 10 major clients of some of the selected firms, as well as about 15 contractors who worked with them, were interviewed to obtain confirmatory data, especially on their level of involvement by the consultancy firms at the design phase of the project delivery process.

FINDINGS

CONSULTANCY FIRMS' FAMILIARITY WITH THE CONCEPT OF LEAN THINKING

It was found that there was some level of awareness among consultants regarding basic principles of managing production systems to minimize waste and maximize value for clients in the project delivery process. There was, however, a gap between the awareness of these basic principles of waste minimization and value maximization on the one hand, and a conscious effort by the firms to establish project delivery systems that focus on waste minimization and value maximization on the otherhand.

The respondents reported a lack of familiarity with the concept of lean thinking. Only 5% of the sample had been involved in the application of lean thinking, 58% claimed to be 'just aware' of it and the remainder of the sample claimed not to be aware of it at all. It is worth noting that even though about 5% of the respondents indicate that they had been involved in the application of the concept of lean thinking, they admitted that no special lean production tool, such as the LPDS, was applied to reduce waste in the design process. Their rationale was that once they try to ensure that waste was reduced, even in the conventional project delivery system, they were applying the concept of lean thinking.

APPLICATION OF THE LPDS

The survey showed that most activities carried out by the consultants at the pre-contract stage are outside the context of the LPDS. Process design at the design phase is a very important activity in the LPDS. However, Figure 3 shows that only 13.2% of the respondents gave an indication that they undertook process design at the design phase of projects. The study also showed that product design was undertaken only at the design phase by most consultants, contrary to a requirement by the LPDS that the product design process be balanced between the design and supply phases. While about 80% indicated that they undertook product design exclusively at the design phase, only about 20% gave an indication that they undertook some form of product design at the supply phase of the project delivery process (Figure 3).

The minimal integration of design and supply activities by Ghanaian firms is further manifested in the indication by 69.1% of the respondents that they undertook detailed engineering only at the design phase while only 20.6% indicated carrying out detailed engineering at the supply stage, as shown in Figure 3. Another significant deviation from the requirements of the LPDS is reflected in the fact that design criteria are not only uncommon in the activities of the consultants, but also mostly concentrated at the design stage by the few who undertake it. The requirement of the LPDS is that design criteria formulation should be carried out at the project definition phase.

Work structuring and production control are required by the LPDS to be carried out across all the phases of the project delivery process to ensure minimization of waste. The survey showed, however, that work structuring and production control are undertaken by very few consultants especially at the project definition, design and supply phases of the project delivery process. Figure 3 shows that only 7.4% and 20.6% of the respondents indicated that at the design stage they undertook work structuring and production control, respectively.

LEVEL OF PARTICIPATION/INVOLVEMENT OF PROJECT PARTICIPANTS IN PRE-CONTRACT ACTIVITIES

According to Harris and McCaffer (2001), achieving good results with lean thinking to reduce waste requires some common tools of production management such as the integration of design and construction, multifunctional task groups and employee participation. All these production management tools require a high level of involvement and participation among such key stakeholders in the project delivery process as consultants, clients, contractors and specialists/suppliers.

Involvement of contractors across the stages of pre-contract activities by consultants was minimal. The highest indication of involvement of contractors by consultants was at the contract documentation stage, which stood at 46.7%. The rest of the indications of involvement of contractors ranged between 6.7% and 13.3%. None of the respondents involved contractors in conceptual design, sketch design and detail design as shown in Figure 4. The indication of involvement of specialists/suppliers,

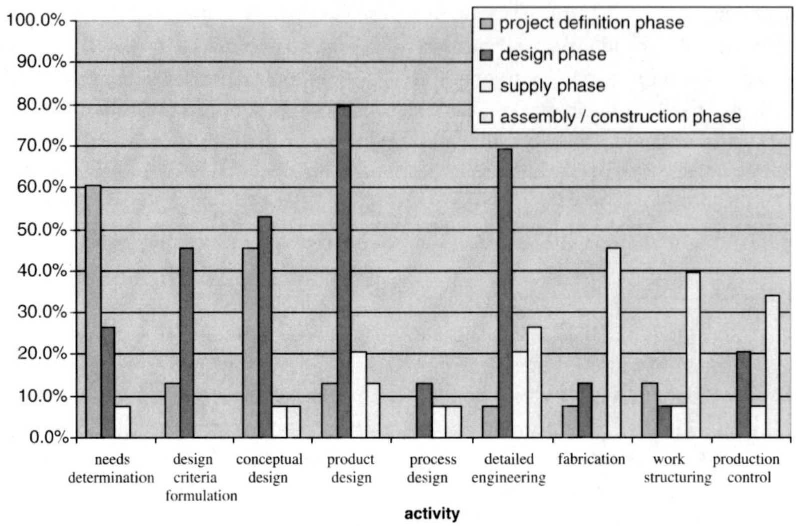

FIGURE 3 Activities undertaken in the project delivery process at various phases

although higher than that of contractors, was also minimal. While 46.7% of the consultants indicated that they sought the participation of specialists/suppliers in scheduling and specification, only 33.3% involved specialists/suppliers at the detail design phase of pre-contract activities.

The dominant role of clients in the choice of site and selection of project participants was reflected in

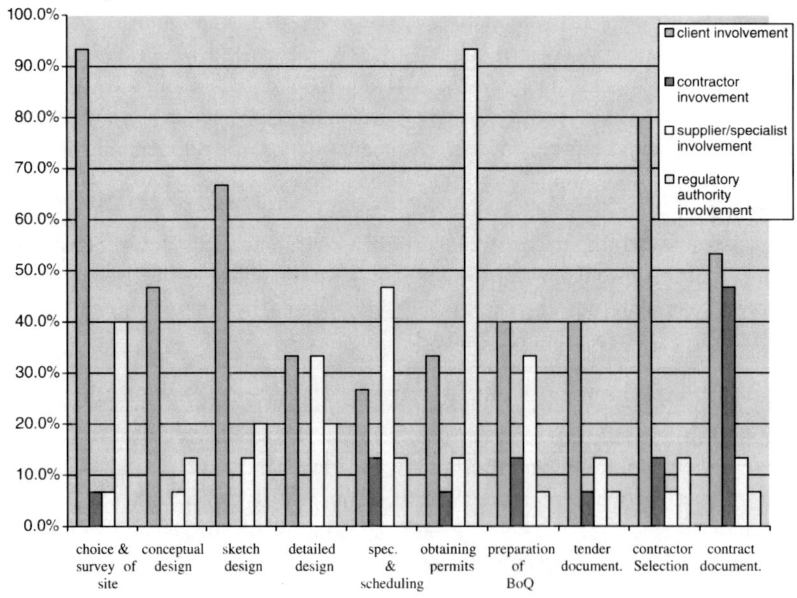

FIGURE 4 Involvement of project participants in pre-contract activities by consultants

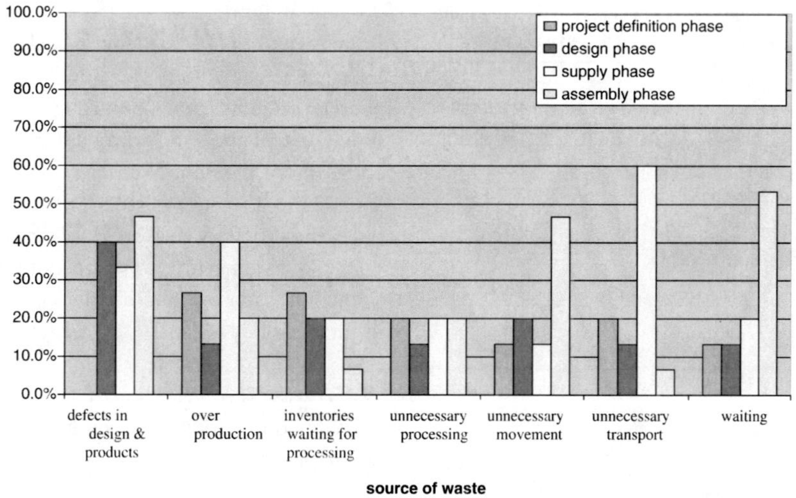

FIGURE 5 Recognition of sources of waste by consultants in the project delivery process

the indication by more than 93% of the respondents that clients participate in the selection of site and 80% indicating that they involve clients in the selection of contractors and suppliers/specialists. The significant influence by regulatory authorities in the acquisition of development permits and certificates, sometimes seen as limitations to continual flow, was confirmed by more than 93% of the consultants (Figure 4).

WASTE IN THE DESIGN PROCESS

One of the important requirements in the practice of lean thinking is the ability to identify and eliminate waste in the project delivery processes. Waste, according to Koskela and Huovila (1997), refers to what is unnecessary for task completion and value generation. Unnecessary or non-value-adding steps are seen to be associated with various design and documentation processes, thus leading to a lot of waste generation.

RECOGNITION OF SOURCES OF WASTE IN THE DESIGN PROCESS BY CONSULTANTS

The survey revealed a low level of recognition by consultants of the sources of waste across the phases of the project delivery process. A small fraction of the consultants associated waste with project definition and design activities. As shown in Figure 5, the highest recognition of waste in the design process was 40%, and this was seen to arise from defects in design. Only 13.3% of the respondents saw waiting as a source of waste that could arise during design and documentation activities. Unnecessary processing was the least recognized source of waste, with only 13.3–20% of respondents associating this source of waste with the various project phases (Figure 5).

CONSULTANTS' UNDERSTANDING OF WASTE

The understanding of waste as that which is unnecessary for task completion was highly recognized. Of the respondents, 73.5% understood waste in design and documentation to be 'that which is unnecessary for task completion and value generation' (Table 2). Understanding waste to be negative iteration and design errors was, however, low among the consultants. While 39.7% of the consultants understood design errors as a form of waste, only 26.5% saw negative iteration as a form of waste in design.

NEGATIVE ITERATION REDUCTION TECHNIQUES

Although negative iteration is one of the major forms of waste in the design process, it was understood by only 26.5% of the respondents (Table 2). It was

TABLE 2 Understanding waste in design and documentation

WASTE IN DESIGN AND DOCUMENTATION	CONSULTANTS	
	$N = 68$	
	NUMBER	PERCENTAGE
That which is unnecessary for task completion and value generation	50	73.5
Negative iteration (non-value-generating design options)	18	26.5
Design errors	27	39.7
Other	0	0
Mean	23.7	34.9

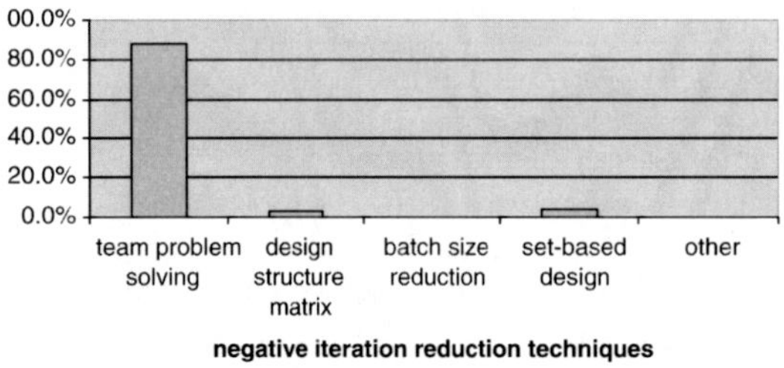

FIGURE 6 Techniques for reducing negative iteration in design

discovered, however, that if negative iteration or non-value-generating design options are understood to be waste, the popular technique of reducing it will be team problem solving. About 88.2% indicated familiarity with the team problem solving technique, while less than 5% showed familiarity with techniques such as DSM and set-based design (Figure 6). No one was familiar with batch size reduction as a technique of reducing negative iteration.

DISCUSSION OF FINDINGS

LEVEL OF FAMILIARITY WITH THE LEAN THINKING CONCEPT

The questionnaire survey and interview showed that there exists a low level of familiarity with the concept of lean thinking as a tool for minimizing waste in the project delivery process. The few consultants who were even aware of the concept of lean thinking had never adopted a special system such as the LPDS to consciously reduce waste and maximize value to clients. The traditional procurement system whereby the design and documentation activities are completely separated from construction activities is still the most widely used system in the project delivery process. Public sector projects are particularly limited by the provisions of the Public Procurement Act of 2003 in an attempt to adopt more integrated systems. Procurement systems such as design and build, even though they fall short of some lean thinking principles, are the common alternatives to the traditional system towards ensuring speedy delivery of projects.

WASTE IN THE PROJECT DELIVERY PROCESS

Even though we found that some level of consciousness regarding the waste associated with project delivery steps exists, design and documentation processes at the pre-construction stage still typify sources of waste such as waiting, delays, defects and inventories. These sources of waste, despite having serious adverse effects on the speedy delivery of value to clients, appear to be too subtle to be tracked by consultants. Responses from consultants during the survey indicated that about 73% and 60% were unaware of negative design iteration and design errors, respectively, as sources of waste. Given that one of the primary activities in the effective application of lean thinking requires identifying and eliminating waste, the inability to recognize waste in design and documentation is a huge setback towards implementing lean thinking principles at the design stage.

INVOLVEMENT OF PROJECT PARTICIPANTS

Management of design and documentation activities at the pre-contract stage falls short of what is required by the tenets of lean thinking to minimize waste and maximize value to clients. The level of involvement of many project participants in the various stages of design and documentation processes is low, thus limiting continual flow in design and documentation, as well as resulting in waiting, delays and unnecessary processing.

The activities of the architect at the design stage, for instance, were found in most cases to be separated from that of the structural engineer and services engineer. The architect would therefore usually complete his designs without the involvement of the structural engineer and services engineer and then give the completed detail architectural designs to them for the structural and services details. This practice will clearly inhibit continual flow in the design process and result in waiting and delays.

PLANNING AND CONTROL OF DESIGN AND DOCUMENTATION PROCESSES

Most of the projects in all the surveyed consultancy firms were designed and documented without reference to any activity programme or work plan specifically drawn for the design and documentation processes. Most of the consultants admitted that even though, as a result of demands from some clients, they tried to provide an activity schedule for some design and documentation activities, it was not part of their usual practice to draw activity schedules and work plans when designs were undertaken. It was therefore not surprising that lean design techniques involving the use of the activity definition model, set-based design and DSM were completely unfamiliar to virtually all the consultants.

CONCLUSION AND RECOMMENDATIONS

The process of building design is associated with some form of waste generation. The waste generated during the building design process is, in most cases, subtle and appears to be difficult to track and manage by consultants in Ghana. The failure of the sample to identify and manage waste in the design process is due to the fact that they still approach project delivery in a linear fashion without recognizing the iterative nature of the construction project delivery process. Negative iteration, which is seen as one of the major forms of waste in the design process, is not well understood or recognized by the consultants who took part in this research. Design errors and unnecessary processing are other forms of design waste that are less recognized as forms of waste. Given that the majority of the sample have not applied lean thinking, it is not surprising that they are also unfamiliar with waste reduction techniques such as DSM, set-based design and batch size reduction.

The procurement process in Ghana, especially for public sector projects, is largely through the traditional procurement route, in which case there is not only less involvement of team members, but also design and construction processes are treated as linear processes and not iterative. Regulations on procurement processes in Ghana should therefore be modified to accommodate more involvement of team members as well as recognize the iterative nature of design to enhance waste management in the design process. This requires a shift towards a lean culture, which will involve promotion, training and education initiatives. For example, there is the need for professional bodies like the Ghana Institute of Architects and the Ghana Institution of Engineers to get members exposed (through continuous professional development programmes) to the concept of lean thinking as well as various waste reduction techniques in the design process.

The inability of most consultants to identify and manage the waste associated with the design process is partly attributable to the fact that design process management and the concept of lean thinking are currently not given much attention in the training of professionals such as architects and engineers in Ghana. There is, therefore, the need for the teaching of the concept of lean thinking to be introduced or strengthened in the academic and professional training of students pursuing construction-related disciplines such as architecture, structural engineering and surveying at both the undergraduate and postgraduate level. This will enhance the familiarity of these trainees with the concept and make them more conscious of managing waste in the building design process after their training.

REFERENCES

Austin, S., Baldwin, A. and Newton, A., 1994, 'Manipulating the flow of design information to improve the programming of building design', *Construction Management and Economics, London, Spon* **12**(5), 445–455.

Austin, S., Baldwin, A., Li, B. and Waskett, P., 1999, 'Analytical design planning technique: a model of the detailed building design process', *Design Studies* **20**(3), 279–296.

Austin, S., Baldwin, A., Li, B. and Waskett, P., 2000, 'Analytical design planning technique (AdePT): a dependency structure matrix tool to schedule the building design process', *Construction Management and Economics* **18**(2), 173–182.

Ballard, G., 2000, 'Positive vs negative iteration in design', paper submitted to the 8th International Conference on Lean Construction, Brighton, 2–7 [available at: www.leanconstruction.org/pdf/05.pdf].

Ballard, G. and Koskela, L., 1998, 'On the agenda of design management research', *Proceedings of the 6th Annual Conference of the International Group for Lean Construction, Guarujá, Brazil*, 13–15 August 1998.

Chua, D.K.H. and Tyagi, A., 2001, 'Process–parameter–interface model for lean design management', *Proceedings of the 9th Annual Conference, International Group for Lean Construction, Singapore*, 1–12.

Coles, E.J., 1990, *Design Management: A Study of Practice in the Building Industry*, The Chartered Institute of Building, Occasional Paper No. 40, 32pp.

Cornick, T., 1991, *Quality Management for Building Design*, Butterworth, Rushden, 218.

Emmitt, S., 2002, *Architectural Technology*, Blackwell Science, Oxford, 37–66.

Ferguson, I., 1989. *Buildability in Practice*, Mitchell, London, 175.

Formoso, C., Tzortzopoulos, P., Jobim, M. and Liedtke, R., 1998, 'Developing a protocol for managing the design process in the building industry', *Proceedings of the 6th Annual Conference, International Group for Lean Construction, Guarujá, Brazil*.

Government of Ghana, 2003, *Public Procurement Act, 2003 (Act 663)*, Ghana Publishing Corporation, Accra, Ghana.

Harris, F. and McCaffer, R., 2001, *Modern Construction Management*, Blackwell Science, UK, 31–32.

Huovila, P., Koskela, L. and Lautanala, M., 1997, 'Fast or concurrent: the art of getting construction improved', in: L.F. Alarcon (ed) *Lean Construction*, A.A. Balkema, Rotterdam, 143–160.

Jones, B. and Riley, M., 1995, 'Collaborative construction agents. Computing in Civil Engineering', *Proceedings of the Second Congress held in Conjunction with A/E/C Systems*.

Josephson, P.-E. and Hammarlund, Y., 1996, Kvalitetsfelkostnader på 90-talet – en studie av sju byggprojekt, Del I Resultat Quality defect costs in the 90s: a study of seven construction projects (in Swedish). Report No. 49. Department of Building Economics and Construction Management, Chalmers University of Technology.

Khedro, T., Genesereth, M.R. and Teicholz, P.M., 1994, 'A framework for collaborative distributed facility engineering', *Computing in Civil Engineering, Proceedings of the First Congress Held in Conjunction with A/E/C Systems*.

Koskela, L., 1992, *Application of the New Production Philosophy to Construction*. Stanford, CIFE, Stanford University. Technical Report No. 72, 35.

Koskela, L., 2004, *Making Do – The Eighth Category of Waste*, International Group for Lean Construction 12, Elsinore, Denmark.

Koskela, L. and Huovila, P., 1997, 'On foundations of concurrent engineering', in: C. Anumba and N. Evbuomwan (eds), *Concurrent Engineering in Construction CEC97*, 3–4 July, The Institution of Structural Engineers, London, 22–32.

Macomber, H. and Howell, G.A., 2004, *Two Great Wastes in Organizations – A Typology for Addressing the Concern for the Underutilization of Human Potential*, International Group for Lean Construction 12, Elsinore, Denmark.

Melhado, S.B., 1994, *Qualidade do projeto na construção de edifícios: aplicação ao caso das empresas de incorporação e construção*, Doctorate Thesis, Escola Politécnica, Universidade de São Paulo.

Melhado, S.B., 1995, *O que é qualidade de projeto?: uma discussão acerca das mudanças conceituais necessárias para a melhoria da qualidade na construção de edificações*. In: *Proc. Workshop Qualidade de Projeto/RS*, Porto Alegre. Porto Alegre, Departamento de Engenharia Civil/Escola Politécnica/PUC-RS.

Melhado, S.B., 1998, 'Designing for lean construction', *Proceedings of International Group for Lean Construction* [available at www.ce.berkeley.edu/~tommelein/IGLC-6/Melhado.pdf].

Mossman, A., Ballard, G. and Pasquire, C., 2010. *Lean project delivery – innovation in integrated design & delivery*, Draft for Architectural Engineering and Design Management Special Issue on Integrated Design & Development Systems (IDDS) [available at http://independent.academia.edu/documents/0095/5229/Mossman].

Nicco-Annan, J., 2006, 'Partnering in construction', *The Quantity Surveyor* **1**, 14–19.

Ohno, T., 1978, *Toyota Production System, Beyond Large-Scale Production*, Productivity Press, Cambridge, MA, 1988.

Reinertsen, D.G., 1997, *Managing the Design Factory*, The Free Press, New York, NY, 269.

Sverlinger, P.-O., 1996, *'Organisatorisk samordning vid projektering: en studie ur ett konsultföretagsperspektiv'*, Lic.Eng. Thesis, Report 44, Department of Building Economics and Construction Management, Chalmers University of Technology, Göteborg (in Swedish).

Tzortzopoulos, P. and Formoso, C.T., 1999, 'Considerations on application of lean construction principles to design management', *Proceedings of the 7th Annual Conference International Group for Lean Construction, Berkeley, CA*, 335–344.

Undurraga, M., 1996, 'Construction productivity and housing financing', *Seminar and Workshop Interamerican Housing Union, Ciudad de Mexico, D.F., Mexico*, 28–29 October.

Westring, G., 1997, *Ghana Public Procurement Reform*. An audit report prepared for the World Bank, Advokatfirman Cederquist KB, Stockholm.

World Bank, 1996, *Ghana 1996 Country Procurement Assessment Report*, The World Bank, Washington, DC.

World Bank, 2003, *Ghana 2003 Country Procurement Assessment Report*, Ghana Country Department, The World Bank, Washington, DC.

An Investigation into the Adoption of Target Costing on Brazilian Public Social Housing Projects

Ana Mitsuko Jacomit and Ariovaldo Denis Granja

Construction Management and Technology Research Group (GTE), Department of Architecture and Building, School of Civil Engineering, Architecture and Urban Design, State University of Campinas, UNICAMP, CP 6021, 13083-852 Campinas, SP, Brazil

Abstract

The suitability of target costing for use in construction remains controversial and specific studies to assist its implementation are lacking. This research aims to critically inspect the contextual characteristics that may influence its implementation during the product development process (PDP) for social housing projects in Brazil. An exploratory and instrumental case study within a state social housing provider company provided PDP-related data. Data collection involved conducting unstructured interviews of higher management and diverse technical staff, and the analysis of company documents and records. A framework describing how target costing is applied in PDP is presented and general propositions are drawn from a theoretical basis built on the relationship between the characteristics of the analysed context and target costing theory. Opportunities and obstacles to target costing applicability were derived from them, as well as clarifications about relevant issues concerning the implementation process of this approach. Although the possibilities of applying target costing to social housing projects are limited by its intrinsic characteristics and traditional institutional barriers to new procurement methods, this article contributes to the discussion for more efficient and proactive ways of conceiving social housing projects, in an effort to deliver more value to stakeholders and end-users.

INTRODUCTION

A new management practice is typically a solution to a practical problem facing industry (Ansari *et al.*, 2007). If the solution is successful, the practice is further developed and it may generate interest from others outside the organization who would like to adopt it or study how it works (Ansari *et al.*, 2007). Target costing, as viewed by authors such as Kato (1993), Ansari and Bell (1997) and Cooper and Slagmulder (1997, 1999), was first developed by Toyota to proactively manage multiyear profit plans that emerged from the organization's strategy. Cost reduction is a means for achieving profit goals (Ansari *et al.*, 2007). Additionally, according to the Japan Accounting Association (1996), it does not intend to manage only costs, but also quality, pricing, functionality, delivery times and other parameters, which are set at the beginning of the product development process (PDP) (Clark and Fujimoto, 1991) at levels that meet the perceived customer needs. Moreover, the achievement of these targets has to be simultaneously attempted in all areas from upstream to downstream processes (Yook *et al.*, 2005).

Although its efficacy has not been tested scientifically, target costing is frequently related to the competitive advantage posed by Japanese auto companies (Sakurai, 1989; Kato, 1993; Tanaka *et al.*, 1993; Ansari and Bell, 1997; Cooper and Slagmulder, 1997). Many other companies, such as Sony, Samsung and Caterpillar, have also implemented this system successfully in the development process of new products (Cooper and Slagmulder, 1997; Ansari *et al.*, 2007). Considering the particular circumstances facing different organizations, questions concerning the factors that influence target costing applicability and success have arisen. Results from an exploratory comparative analysis at six Japanese manufacturing firms indicated that they are influenced by five factors: intensity of competition, type of customer, product characteristics, product strategy and supplier-base strategy (Cooper and Slagmulder, 1997). The study was based on the identification of contextual characteristics related to each factor and the analysis of their impact on the target costing system at each company.

But what factors would arise if Cooper and Slagmulder's research was carried out in a different context, such as construction? What would be the impact of target costing applicability on construction peculiarities, such as one-of-a-kindness, site production and temporary organizations, which frequently lead to complexity, variability and a lack of transparency (Koskela, 2000)? This is the starting point of this research, which aims to critically inspect the characteristics that could have an influence on an implementation of target costing on the development processes of Brazilian social housing projects (SHP).

The Brazilian housing deficit is around 5.5 million dwellings,[1] for a population of 192 million inhabitants.[2] The growing demand for social housing has increased pressure on public agencies for the construction of new housing units. A process of standardization, recognized as an inherent feature of the industry (Ball, 1983), was identified as the best way to deal with this demand. Production in volume develops economies of scale that can produce standardized products (Nguo and Armitage, 2008), which could theoretically minimize the undesirable effects of construction peculiarities, especially 'one-of-a-kindness' (Koskela, 2000).

When there is a limited budget for the provision of a housing project, the less that is spent on each unit means a greater number of units can be built. A fact worth noting is that Brazilian SHP are usually known for poor quality and functionality. However, regardless of the amount spent on each unit, the question remains: Is it possible to deliver more quality and functionality for the same cost?

Due to the profitless nature of the analysed company, the target costing system is envisioned here as a trigger of SHP PDP transformation, aiming to increase value by improving product functionality without raising costs – which is in agreement with the reasoning of Ansari *et al.* (2007) about target costing applicability for non-profit organizations. The analysed company is the main provider of social housing in Brazil – which has not yet applied target costing. It is a state company in charge of managing the PDP and production process, adopting a design-bid-build method for the delivery and commercialization of its units.

However, how can one be sure that the target costing approach is as effective in the social housing context, or even in construction as a whole, as it is in manufacturing? In fact, the motivation for this study originated from the diversity of opinions about the applicability of target costing in construction (Jacomit *et al.*, 2008). While Nicolini *et al.* (2000) concluded that it cannot be applied to construction due to industry procurement practices, Ballard and Reiser (2004) and Robert and Granja (2006) reported successful implementations. They, however, clearly did not follow the same approach. Nicolini *et al.* (2000) remained loyal to the manufacturing theory whereas the other authors applied adapted approaches to construction.

Thus, this article presents a tentative framework that intends to demonstrate how the target costing system can be applied in association with the PDP, mainly, to the six key principles established by the International Consortium for Advanced Management: price-led costing, customer focus, design centred, cross-functional teams, value-chain involvement and life-cycle orientation (Ansari and Bell, 1997). After identifying what is understood as target costing, it was confronted with the actual PDP of SHP in order to identify contextual characteristics that would (i) raise and (ii) reduce target costing applicability or (iii) neither raise nor reduce it but, rather, influence it by directing its implementation process. Characteristics (i) and (ii) were selected, respectively, due to their similarities to and differences from the general characteristics of manufacturing. Characteristics (iii) were selected based on their similarities to or differences from target costing approaches adopted from past target costing implementations.

As the analysed company did not apply target costing, the relationship between its context and target costing was established through a tentative causal model that describes how each chosen context characteristic would affect the target costing applicability. Thus, it is supposed to work as a bridge between the SHP PDP and the target costing theory, anticipating likely implementation opportunities and obstacles.

ORIGINS OF TARGET COSTING

For many authors, the origins of target costing come from the Japanese car industry in the early 1960s (Monden, 1995; Nicolini *et al.*, 2000; Yook *et al.*, 2005; Okano and Suzuki, 2007). However, there is anecdotal evidence of the adoption of related concepts by Ford at the beginning of the 20th century, such as the establishment of a target price to guide the development process of the Model T (Monden and Sakurai, 1989; Cooper and Slagmulder, 1997, Feil *et al.*, 2004). Probably, this evidence came from an interpretation of the following statement: 'In our own production, we set ourselves tasks – sometimes we arbitrarily fix prices and then invariably we are able to make them; whereas, if we merely accepted things as they are, we should never get anywhere' (Ford and Crowther, 1926). There is also evidence of the adoption of concepts related to target costing in the Volkswagen Beetle development process in the 1930s (Feil *et al.*, 2004). Thus, it is apparent that Japanese corporations incorporated Western practices such as value engineering (Miles, 1989), which became the roots of target costing as we know it today (Monden and Sakurai, 1989; Ansari *et al.*, 2007). They did the same with Deming's approach to quality management (Deming, 1982), which was further developed into total quality management, overwhelmingly adopted by Japanese companies today.

According to Cooper (2002), the Japanese auto industry was responsible for elevating target costing from a simple cost reduction exercise into a strategic system for managing profits. However, the initial concepts applied by Ford seem to have evolved in the USA into another, although similar theory, known as 'design to cost' (Everaert *et al.*, 2006). This theory has its roots in the Department of Defense (DoD) and is strongly related to the development of the American arms industry in the 1960s (Michaels and Wood, 1989; Morris, 1994). Yoshikawa *et al.* (1993) affirms that the US strategy is focused on the internal capabilities of an organization, whereas the Japanese strategy is externally oriented, market focused. In fact, in the target costing approach, the PDP is oriented by the market, with the definition of the allowable cost, as well as the minimum standards for quality and functionality, based on customers' expectations (Cooper and Slagmulder, 1997). In the design to cost approach, the PDP is entirely guided by a single client (Michaels and Wood, 1989). In the case described by Michaels and Wood (1989), the DoD established the project 'affordable budget', from which the project team selected the best available options in terms of schedule, functionality and quality.

Additionally, the emphasis on life-cycle costing under the two approaches assumes different points of view. Design to cost emphasizes its implications to the final user or client (DoD), that is, considering operations and maintenance costs. But, in the target costing literature (Monden, 1995; Cooper and Slagmulder, 1997, 1999), life-cycle costing is frequently related to producer profit management, that is, considering all costs and revenues that occur during the manufacturing life cycle, until production is stopped. Evidently, these differences are the result of contextual differences: the Japanese approach was first developed to raise a company's competitiveness globally, whereas the American approach was directed at reducing the maintenance and operational costs of weapons to the government.

However, Ansari *et al.* (1999) argue that the strategy being applied by the DoD is, in fact, derived from Japanese target costing and is denominated by the DoD as 'cost as an independent variable' (CAIV), not acknowledging the expression 'design to cost'. They reported that CAIV is applied to products such as fighter aircraft, where congressional budget limits represent the target cost and fighter specifications – such as safety, manoeuvrability and bomb payloads – serving as customer requirements to guide PDP (Ansari *et al.*, 1999, 2007).

Target costing itself also receives different denominations. In early publications, it was called *genka kikaku*, cost planning and cost projection systems (Kato, 1993). Dekker and Smidt (2003) found further synonymous terms in use such as basic net price, manufacturing cost reduction, pre-calculation and direct cost feasibility study.

In the construction industry, especially in the building sector, some authors seem to adopt both – American and Japanese – strategies in combination. For example, Ballard and Reiser (2004), used a strategy called 'design to target cost' inserted in the broader 'lean project delivery system', based on the Toyota Product Development System (Ballard, 2008). The construction project analysed by Ballard and Reiser (2004) is characterized by a single client interested in reduced operational and maintenance costs and increased value – a context that resembles the original context of the design to cost approach.

Even when considering only the theory on target costing, one can find different definitions for the most basic terms, such as the target cost. According to Monden (1995), Ansari *et al.* (2006, 2007) and Ellram (2006), the target cost of each product is determined by the difference between the market price and the profit per unit. But, according to Cooper and Slagmulder (1997, 1999), Ballard and Reiser (2004), Dekker and Smidt (2003) and Nicolini *et al.* (2000), this is the definition of the allowable cost. Therefore, the target cost is determined based on the allowable cost and the capabilities of a company's production department to reduce costs. It is often established at an intermediate value between the estimated manufacturing cost – or drifting cost (DC) – and the allowable cost (Cooper and Slagmulder, 1997, 1999; Feil *et al.*, 2004). However, Ballard (2008) argues that the target cost should be set even below the allowable cost, so as to stress the system and push towards innovation.

TARGET COSTING IN THE PDP

As described in the previous section, there are different interpretations of some concepts related to target costing and even of target costing itself. Consequently, before investigating the applicability of the system, it may be necessary to define what is understood here as target costing. Based on Jacomit *et al.* (2008) and following the key principles of target costing (Ansari *et al.*, 1997, 2006), a tentative framework that intends to demonstrate how the target costing system could be applied in association with the PDP of a construction product is presented in Figure 1. This framework was not tested and is mainly the result of a literature review covering both manufacturing and construction contexts.

Steps 1–3 (Figure 1) are related to the principle 'customer focus', in which 'product design is shaped continuously by the voice of the customer and enhancements of product features take place only when they meet customer requirements and customers are willing to pay for them' (Ansari *et al.*, 1997). It could not be different since a customer's perceived value of a product is more important than an objective fact (McNair *et al.*, 2006). Market research plays an important role in this stage of manufacturing.

Steps 4–10 illustrate target cost (TC) determination, and reflect the principle 'price led costing', that is, cost is a function of a market-determined price (Ansari *et al.*, 1997). At this stage, market intelligence and strategic information systems are usually employed. Here, market intelligence comprehends all forms of collecting and processing relevant market data. In target costing literature, strategic information systems usually refer to a special kind of cost table capable of converting value or product characteristics into price (Kato, 1993; Nicolini *et al.*, 2000). As can be inferred from Figure 1, there are different types of cost tables and their functionality is related to 'when' and 'where' they are used (Yoshikawa *et al.*, 1990; Agndal and Nilsson, 2009).

Steps 10–17 describe the determination of the first estimate of the DC, before design starts, and TC subdivision. They are related to the principle 'design

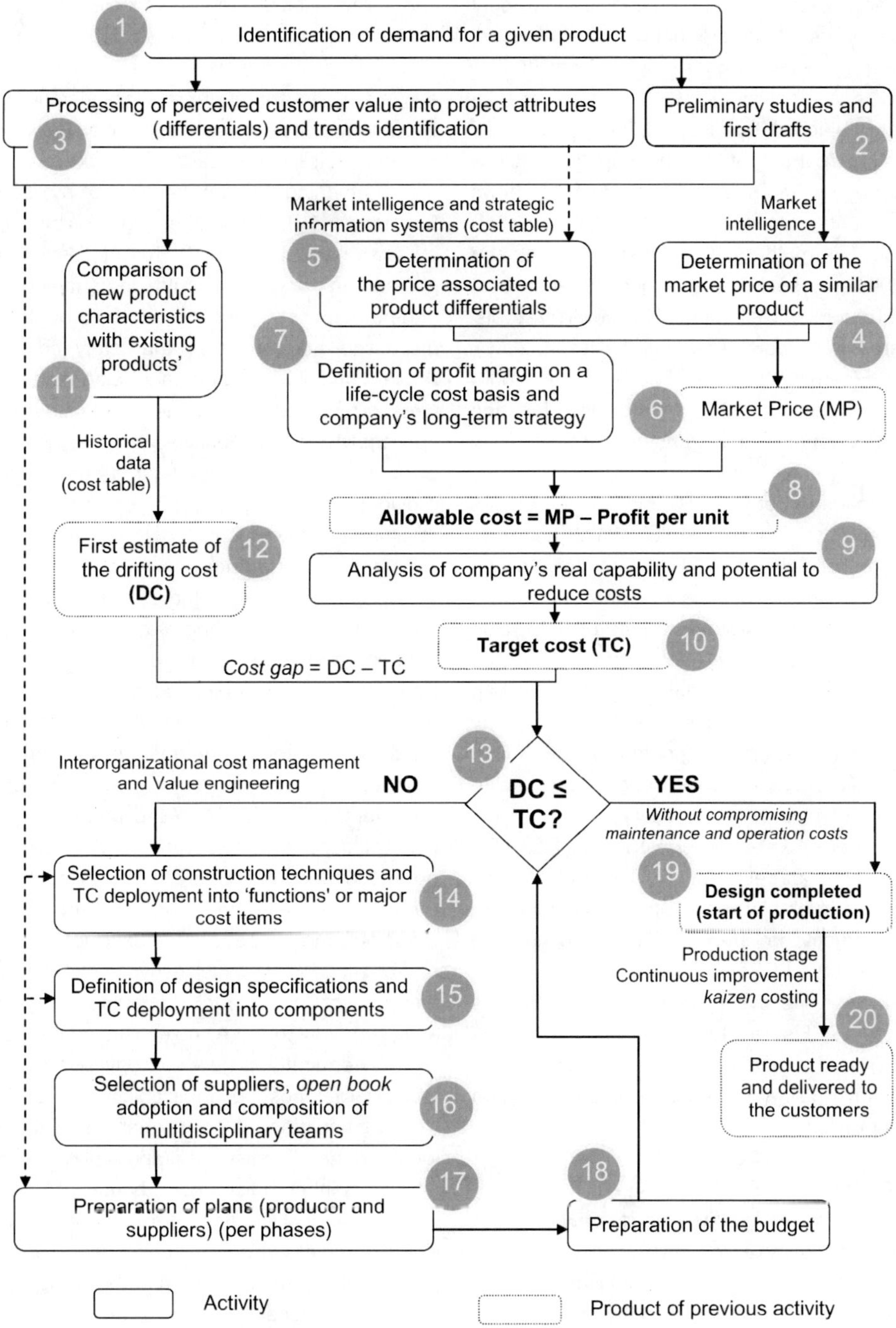

FIGURE 1 Target costing integrated in the PDP

centred', that is, the cost of designing a product before committing it to production (Ansari *et al.*, 1997). Inter-organizational cost management (IOCM) assists TC by subdividing TC into functions or components, which determine, respectively, the TC of designing parts for each supplier. IOCM represents buyers' and suppliers' coordinated efforts to reduce costs (Cooper and Yoshikawa, 1994; Cooper and Slagmulder, 1999, 2004; Agndal and Nilsson, 2009), by seeking a balance between cost control and supplier well-being. The auxiliary 'tools' of IOCM are a cost table (Agndal and Nilsson, 2009), open book (Ellram, 1996, 2000; Nicolini *et al.*, 2000; Agndal and Nilsson, 2009) and quality function deployment (Cooper and Slagmulder, 1997; Ansari *et al.*, 2007).

The comparison between TC and DC determines the cost gap (cost gap = DC − TC), which needs to be null before allowing the product to be released for production. While the main function of IOCM is to enable the realistic establishment of TC, value engineering tools such as workshops and value analysis play a major role in annulling the cost gap (without compromising quality and functionality levels established at the beginning of the PDP), since the main goal of value engineering is to identify and eliminate unnecessary costs (Miles, 1989). Thus, steps 13–18 are repeated until DC ≤ TC.

Although the target costing principles 'cross-functional teams' and 'value-chain involvement' are addressed directly in step 16, they do in fact permeate the entire PDP.

The principle 'life cycle orientation' can directly influence step 7 (Figure 1), which is related to the definition of profit margin, or the whole process, depending on the target costing implementation context, as previously discussed.

RESEARCH METHOD

According to Ansari *et al.* (2007), the majority of topics in a research line vary according to the maturity of the line, starting from conceptualization and logical inference of constructs and hypothesis, and progressing to the test of constructs and propositions. These authors concluded that the research line in which target costing is included is still in the early stages of maturing. This means that testing constructs and relationships among variables are particularly challenging because, in many cases, constructs and/or variables are not well known. On this basis, this article aims to contribute to the generation of theory about target costing applicability in a specific context.

The main theoretical frameworks employed are the intervening factors developed by Cooper and Slagmulder (1997) and the target costing principles by Ansari and Bell (1997). Target costing principles played an important role in shaping what is understood as target costing, although it is not discussed here whether they truly correspond to those principles. Cooper and Slagmulder's factors were employed aiming to study target costing applicability in the analysed context. But, although Cooper and Slagmulder's factors and characteristics provided the starting point for this research, the resulting characteristics do not resemble those chosen by those authors.

Cooper and Slagmulder's factors determine the ideal context for applying target costing, that is, the context from where more benefits can be derived from it. For example, according to them, the greater the variety of products in a line, the more benefits it could render, since it increases product development priority and further justifies the allocation of funds to the budget. However, in construction, a small variety of products being produced and replicated could, in fact, increase a product's development priority and its budget, reducing the impact of construction peculiarity one-of-a-kindness. Thus, early results based on the analysis of Cooper and Slagmulder's factors in the analysed context were subjective and not conclusive.

Consequently, a new strategy was elaborated to study target costing applicability. It was based on the identification of contextual characteristics that would either (i) raise its applicability, (ii) reduce it or (iii) neither raise nor reduce it but, rather, influence it, by directing its implementation process. Characteristics (i) were those that resemble the general characteristics of manufacturing and whose causal relationship with target costing was identified. Characteristics (ii) were those that differ from manufacturing general characteristics and whose causal relationship with target costing was also identified.

Characteristics (iii) were selected based on either their similarities to or differences from target costing approaches adopted in past target costing implementations. Thus, a causal model was elaborated in a tentative way to explain why, in some target costing implementations in construction, (a) the allowable cost is not set-based on price, (b) life-cycle costing encompasses costs related to the ownership of the product (i.e. maintenance and operation) rather than those incurred during product development and production, and (c) target costing and PDP assume an internal instead of an external focus.

As it was not possible to find any company applying target costing in the analysed context, target costing and PDP are studied separately and the relationship between them is established on a theoretical basis. Hence, this research can be divided into three main parts: (1) defining what is understood as target costing for the sake of this research, (2) identifying (i)–(iii) and (3) elaborating a causal relationship between (1) and (2). Parts (2) and (3) were developed in parallel.

Part (1) was based on a literature review and its main product is a framework that describes how target costing could be applied to PDP, which was introduced in the previous section.

Part (2) comprised an exploratory and instrumental case study carried out at the main supplier of social housing in Brazil, for a period of 1 year. It can be seen as an exploratory case study because the collection of data was undertaken prior to the final definition of the study's questions and because, sometimes, intuitive paths were followed (Yin, 2003). It is instrumental because the examined case was chosen primarily to provide insights into an issue; that is, it is of secondary interest and plays the supportive role of facilitating the understanding of something else (Stake, 2005). The techniques of data collection consisted of unstructured interviews with the top quality manager and diverse technical staff (at least five interviews, including one by e-mail), field observation and the analysis of documents and records, such as budgets, plans, contracts, material and labour specifications and bills of quantities.

The analysed company is the Housing Authority of the State of São Paulo and the largest developer of social housing since 1986 in the state. It is in charge of managing SHP development and production processes as well as commercialization of the units. Company policies are directed to satisfy the housing needs of low-income level families, through subsidies and governmental funding. The analysed product can be represented by a standard design model, based on repetition of identical residential units.

In part (3), tentative general propositions (Flyvbjerg, 2006) were drawn aiming to explain how characteristics (i)–(iii) identified in part (2) influence target costing, represented by the framework developed in part (1). These propositions are based on a comparative analysis (Strauss, 1987) of evidence collected during the case study and information found in the literature.

SHP PDP

The analysed company has developed and/or financed 439,161 housing units.[3] Among residents, family incomes range from 0 to 10 SM (1 SM corresponded to approximately US$288 in September 2010), with the highest concentration between 1 and 4 SM. A unit's price corresponds to the transfer of construction, administrative, financial and other costs, proportionally – profit margin does not exist (Jacomit and Granja, 2010). Payment is amortized over 25 years. Each family pays a small monthly instalment, which varies according to the family's income, ranging from 15 to 30% of it (Jacomit and Granja, 2010). The remainder of the debt is covered by the state. Each housing unit has 38 m^2 and cost about US$20,000 to construct, in 2003.

The location of the projects reflects a lack of sufficient funds, which is why they are built on the outskirts of cities. A typical project corresponds to the repetition of a standard product inside a designed area. The standard product is a five-storey building without an elevator, as shown in Figure 2. Buildings are arranged in an orthogonal disposition so that they form an 'H' (Granja *et al.*, 2009).

The implementation of a social housing programme begins with the identification of a regional demand and the definition of an adequate housing solution that will meet this demand (Jacomit and Granja, 2010). It is characterized by families living in the municipality, who do not own a house or apartment and whose income ranges from 1 to 10 SM. Generally, the city

FIGURE 2 Typical SHP developed by the analysed company
Source: Granja *et al.* (2009).

government is involved and donates the land and urban infrastructure. Then, the housing company stipulates the number of units to be built, their construction schedule, the project's basic premises and other parameters that will make up the term of reference for procurement of design and construction services. During the construction phase, the housing company registers the prospective families and sells the units.

As the supply is not equal to demand, the future residents are chosen by lottery.

INTERVENING CHARACTERISTICS

Target costing is commonly applied by automobile and other assembly line manufacturers (Ansari *et al.*, 2007). A simplified stakeholders' arrangement consists of a big company that manages the development and production process and runs the target costing system and its suppliers, who are subordinated to it. In construction, a stakeholders' arrangement or 'project coalition' (Winch, 2002) varies widely. As described by Winch (2002), it can be integrated, separated or a coalition, being design-build, design-bid-build and turn-key, as some examples. As hierarchical positions vary according to each project's arrangement, the definition of which company is going to apply target costing and the purpose of this implementation precedes and shapes how to do it.

Strategies followed exploratory target costing implementations in construction (Nicolini *et al.*, 2000; Ballard and Reiser, 2004; Robert and Granja, 2006) and were directed to the fulfilment of customer needs, as if the client were applying the target costing himself. However, major decisions – such as defining target costs – were made by the main contractor in collaboration with other stakeholders. Robert and Granja (2006) adopted incentive contracts to stimulate collaboration, but just with one subcontractor. The case described by Nicolini *et al.* (2000) was not successful due to insufficient collaboration resulting from a conflict of interest among the stakeholders. Ballard and Reiser (2004) do not make it clear why the main contractor and subcontractors should work collaboratively towards the client's interests.

In this study, it was assumed that the company applying target costing – as described in Figure 1 – corresponded to the analysed company in its current state. The purpose of the implementation was to increase the value added to SHP developed by the company, by improving the products' functionality without raising costs. Considering this scenario, theoretical opportunities and obstacles to target costing implementation were found. Additionally, tentative general propositions dealing with target costing relevant issues, which can shape its implementation, were drawn.

CHARACTERISTICS THAT INCREASE TARGET COSTING APPLICABILITY

Table 1 summarizes SHP characteristics that would raise target costing applicability: how they would influence target costing and which activities would be affected, considering the framework shown in Figure 1.

DESIGN STANDARDIZATION AND REPLICATION OF DESIGN

One of the biggest issues regarding target costing implementation in construction is related to the high level of uncertainty in the early stages of the project, which leads to difficulty in the establishment of a realistic and, at the same time, challenging target

TABLE 1 SHP characteristics that raise target costing applicability

	GENERAL PROPOSITION	INFLUENCE OVER TARGET COSTING FRAMEWORK (FIGURE 1)
Design standardization	It diminishes consideration of individual users' requirements through project design, increasing market research importance	Steps '1 and 3' (capture of customer requirements)
	It facilitates determination of the market price, allowable cost and first estimate of the DC	Steps '4–8' (allowable cost determination), '9 and 10' (target cost determination) and '11 and 12' (DC determination)
Replication of design	It enhances the importance of the PDP and gives an opportunity to designers to improve an existent design based on feedback from occupiers	All steps

cost, which will lead the PDP. The development of standardized solutions or a typology (Kelbaugh, 2007), for instance, provides a portfolio of proven designs, which in turn enables developers to make reasonably accurate estimates of costs, price and production time without delay (Leopold and Bishop, 1983). However, efficient cost feedback requires great detailing efforts in the design phase, and this results in more work in the early stages of design compared with projects not aligned with the target costing approach (Jørgensen and Emmitt, 2009).

Typologies can also hold back another possible counter-argument against target costing implementation in construction, related to the exclusivity of each project (Jacomit and Granja, 2010). Sceptics could argue that additional design fees could supplant earnings from design improvement. However, replication amplifies total earnings related to a design, increasing the likelihood of profits.

As a typology is developed to satisfy families with different profiles, market research and post-occupancy evaluation (POE) are important to take care of their individual needs (Jacomit and Granja, 2010), which may lead, for example, to the adoption of constructive solutions that allow for product differentiation such as layout modification. POE studies can avoid the propagation of design inaccuracies and underlie the design improvement process (Kowaltowski *et al.*, 2006).

Typologies are usually developed based on past references and under the premise that human nature, needs and bodies have not changed, nor have climate or geography, significantly (Kelbaugh, 2007). However, this does not necessarily restrict innovation since designers 'may admit that a design problem presents unprecedented social issues and new technological opportunities' (Kelbaugh, 2007, p.87), providing an opportunity for the application of value engineering. On the other hand, economies of scale invariably lead to some resistance to design changes, which are often limited to internal changes (Nguo and Armitage, 2008).

Standardization in the production of dwellings also acknowledges the benefits of the limited skills of the labour force (Carmona, 2001) and its prior knowledge in carrying out repetitive tasks, which results in lower production costs (Ball, 1983). Additionally, Ball (1983, p.167) argues that 'handicraft assembly of "traditional" but standardized and partly pre-assembled building components' has survived not because the industry 'is backward and conservative, but rather because traditional building methods are the most profitable'.

CHARACTERISTICS THAT REDUCE TARGET COSTING APPLICABILITY

Table 2 summarizes SHP characteristics that would reduce target costing applicability: how they would influence target costing and which activities would be affected, considering the framework shown in Figure 1.

BIDDING PROCESS

The arrangement between stakeholders can interfere in a target costing implementation, mainly if it corresponds

TABLE 2 SHP characteristics that reduce target costing applicability

	GENERAL PROPOSITION	INFLUENCE OVER TARGET COSTING FRAMEWORK (FIGURE 1)
Bidding process	It leads to DC definition based on unrealistic (average) data It prevents contractor collaboration at the beginning of the PDP (which can compromise constructability and cost reduction) It reduces cost reduction opportunities associated with the supply chain It tends to prevent cost reduction through the production phase and enhance transactional costs	Steps '14 and 15' (TC subdivision), 16 (composition of multidisciplinary teams), 17 (preparation of plans) and 18 (preparation of the budget)
Outsourced design	It raises the complexity of value analysis and trade-offs between costs, quality and functionality	Information flow between steps '1 and 3' (capture of customer requirements) and steps '14–17', as well as steps '14–17'

to design-bid-build, which is usually associated with poor coordination and high levels of adversarial relations (Nicolini *et al.*, 2000; Winch, 2002). In fact, those interferences led authors such as Sobotka and Czarnigowska (2007) to infer that target costing cannot be applied in association with design-bid-build.

In the analysed context, characterized by a low-bid public bidding process, these interferences are related to the fact that the contractor and his suppliers are unknown during design development and budgeting. Thus, first, they cannot collaborate with design to improve constructability or compromise with cost targets and, second, the DC cannot be based on actual data applied by the contractor.

Target costing practitioners such as Honda do not simply force less-efficient suppliers to operate at determined price levels regardless of suppliers' profitability (Sisodia *et al.*, 2007). Instead, they work with suppliers to identify inefficiencies that might be preventing them from meeting their target costs. Thus, their target costs are achieved as the supplier gains in efficiency (Sisodia *et al.*, 2007). However, it is easier for large manufacturing companies to count on the collaboration of suppliers than it is for average construction companies to do so.

Winch (2002) describes the adversarial culture that prevails in the industry as: attempts by clients to reduce the scope for opportunism (low bid and over engineering) who have the tendency to generate vicious circles of adversarial behaviour between the parties, which in turn triggers claims and control measures elevating transactional costs, while the production costs (direct and overhead) seem to be pulled down. An attempt to reverse these dynamics is by incentive contracts (Winch, 2002), which intend to balance the cost of risk bearing against the incentive gains that result, that is, to eliminate opportunistic behaviour from information asymmetry and uncertainty. However, although the association of target costing and incentive contracts has been discussed by others such as Robert and Granja (2006), Sobotka and Czarnigowska (2007) and Ballard (2008), no conclusive results have been reported up to now.

OUTSOURCED DESIGN

According to Cooper and Slagmulder (1997, 1999), a product can be defined by three basic dimensions – quality, functionality and cost/price – that compose the 'survival triplet' of a product. Based on customers' requirements and the company's strategy, minimum and maximum parameters are established for each dimension, which determine the 'survival zone'. Those authors argue that 'only products with values within each of these three dimensions [inside the survival zone's borders] stand a chance of being successful' (Cooper and Slagmulder, 1997, p.5). Then, as dimensions are interrelated, the key to profitability is the balance between them. Cooper and Yoshikawa (1994) refer to this balance as 'quality-function-price trade-off'.

Since most of this balancing occurs during design and if design is outsourced, these dynamics need to be described in the contract, which raises target costing complexity. In fact, Ballard (2008) questions the 'feasibility of developing cost models directly from the voice of the customer in the project definition phase', since customers tend to give equal or more importance to intangible factors such as 'external beauty, internal comfort and "suitable to live in", than tangible ones' (Pennanen and Ballard, 2008).

Collaboration between cost estimators (from the housing company) and designers (from the contractor) as well as the adoption of cost modelling systems could reduce the influence of outsourcing design. However, cost models such as building information modelling are still under development and their association with target costing needs further development.

CHARACTERISTICS THAT DIRECT TARGET COSTING IMPLEMENTATION

Table 3 summarizes SHP characteristics that would direct how target costing is implemented: how they would influence target costing and which activities would be affected, considering the framework shown in Figure 1.

IMPLEMENTATION PURPOSE

As discussed in the section 'Origins of target costing', target costing can be applied to many purposes. If the purpose of implementation is related to profit from product sale, as is the case of most (if not all) implementations in the manufacturing context, the

TABLE 3 Characteristics that direct target costing implementation

	GENERAL PROPOSITION	INFLUENCE OVER TARGET COSTING FRAMEWORK (FIGURE 1)
Implementation purpose	*If it is related to profit from product sale:* The allowable cost is based on price and unit profit	Steps '4–8' (allowable cost determination)
	**If it is not related to profit from product sale*:* The allowable cost may not be based on price and unit profit	
Decision maker	*If the decision maker (Bartlett and Howard, 2000) corresponds to 'owner-occupier' or tenant:* Life-cycle costing encompasses costs related to ownership of the product, i.e. maintenance and operation	All steps
	**If the decision maker (Bartlett and Howard, 2000) corresponds to developer or investor*:* Life-cycle costing encompasses costs incurred during product development and production. Costs related to the ownership of the product are raised or reduced according to the decision maker's strategy	Step 7 (profit margin definition)
End-user participation in design	*If they participate in design:* Target costing takes an internal focus	Steps '1 and 3' (capture of customer requirements)
	**If they do not participate in design*:* Target costing takes an external focus	

allowable cost is based on price and unit profit. However, if it is not related to profit from product sale, the allowable cost does not necessarily need to be based on price, as was the case of the implementations described by Michaels and Wood (1989) (budget constraints), Nicolini *et al.* (2000) (historical data), Ballard and Reiser (2004) (budget constraints) and Robert and Granja (2006) (historical data).

In the case of the analysed company, the allowable cost could be determined based on government budget constraints since these constraints could allow for the provision of satisfactory levels of quality and functionality for SHP.

DECISION MAKER

According to Bartlett and Howard (2000), the decision makers are those who decide on the construction of a building, corresponding to the 'owner-occupier' (i.e. those that own a property and occupy it), the tenant, the developer and the investor. The owner-occupier and the tenant are often the only stakeholders with a financial interest in the building during its service life. Thus, if target costing was applied to meet their expectations, life-cycle costing would encompass costs related to the ownership of the product, such as maintenance and operational costs.

But if target costing was applied to meet the expectations of the developer or the investor, life-cycle costing would encompass just costs incurred during project development and construction. Costs related to the ownership of the product would be raised or reduced according to the decision maker's business requirements. In competitive contexts, the consumer market regulates these dynamics, pressing companies to deliver increasingly cost-effective products in terms of acquisition and maintenance/operational costs (McNair *et al.*, 2006).

In the analysed context, considering that the consumer market is composed of low-income families, the competition is low. In the range from 0 to 3 SM, it does not exist, which therefore reduces the users' freedom of choice. In such a context, minimum levels of quality and functionality for SHP should be institutionalized in order to ensure satisfactory 'habitability'. Moreover, the allowable cost should be based on these standards and not the other way around – which would result in poor quality and functionality, considering the strict budget constraints in place.

END USERS' PARTICIPATION IN DESIGN

One of the target costing principles is 'customer focus', which means that new products are developed to meet 'customer price and quality requirements' as opposed to a 'cost-driven development of products', in which a product is pushed on to customers in the hope that they buy it (Ansari *et al.*, 2007). For most manufactured products, it is achieved through the development of an external focus, which means that the PDP team take their 'cue from the market and customers' (Ansari *et al.*, 2007, p.1).

But, when 'customers' take part in the PDP or when the decision maker corresponds to owner-occupier, 'customer focus' does not imply an external focus. This is the case described by Michaels and Wood (1989), Ballard and Reiser (2004) and Nicolini *et al.* (2000).

In the analysed company, end-users do not participate in design. In such cases, POE is important to ensure that SHP are truly developed to meet users' expectations. However, if their expectations are low, as in the case of most SHP users, POE usually indicates high levels of satisfaction despite the poor quality of the constructions (Kowaltowski *et al.*, 2006).

CONCLUSIONS

This article contributes to the promotion of discussions on more efficient ways of developing products in construction. It investigates target costing applicability for SHP by a housing company. However, the target costing framework and propositions about the relationship between target costing and SHP characteristics can be generalized to other construction products, since they possess one or more of those characteristics. The target costing framework describes how this system could be applied in PDP, not considering the influence of contextual peculiarities. SHP characteristics were analysed in order to determine which of them would correspond to obstacles or opportunities to target costing implementation. It was concluded that the standardization and replication of design can be seen as opportunities, and the outsourcing of design and the bidding process as obstacles.

Additionally, some characteristics were identified that would influence how target costing is applied, concerning the definition of allowable cost, life-cycle costing orientation and product development focus. The allowable cost can be based on price or not, depending on the purpose of implementation. Life-cycle costing can focus on costs related to the ownership of a product or those incurred during product development and production, depending on who the decision maker is. Product development can have an internal or external focus, depending on whether or not end-users participated in design.

General propositions were established based on logical and theoretical assumptions and, for instance, the need for further development through more positivist approaches, such as explanatory case studies focusing on, for example, (i) the effect of incentive contracts on suppliers' collaboration in a target costing implementation in construction, (ii) the possibility of applying target costing in association with design-bid-build and (iii) the feasibility of applying target costing with life-cycle costing encompassing all operations and maintenance costs.

The results of this research suggest that target costing could be more successful in an SHP context if applied by a company coordinating PDP and construction, for example adopting the design-build delivery system, which is the case of projects developed through the recently launched Brazilian government housing programme *Minha casa, minha vida* – My home, My life. This programme can also promote an increase in competition based on merits in this sector, which in turn can lead to an increase in SHP quality and value delivery to end-users and clients. Thus, new avenues for research on target costing applicability in SHP by a public or private company coordinating design-build coalitions could also be further explored in similar international contexts.

ACKNOWLEDGEMENTS

The authors thank FINEP (Brazilian Innovation Agency – Research and Projects Financing) for supporting this research, CAPES (Brazilian Federal Agency for Support and Evaluation of Graduate Education) for granting the research fellowship and James R. Jernigan for proofreading the manuscript.

NOTES

1 Estimate conducted by the Brazilian Ministry of Cities, based on data from 2008, João Pinheiro Foundation, Belo Horizonte (MG).

2 Estimate conducted by the Brazilian Institute of Geography and Statistics (IBGE). Available at http://www.ibge.gov.br/english/

3 Data provided by the analysed company in December 2008.

REFERENCES

Agndal, H. and Nilsson, U., 2009, 'Interorganizational cost management in the exchange process', *Management Accounting Research* **20**(2), 85–101.

Ansari, S. and Bell, J., The CAM-I Target Costing Group, 1997, *Target Costing – the Next Frontier in Strategic Cost Management*, Irwin, Chicago, IL.

Ansari, S., Braxton, P. and Coleman, R., 1999, 'Processes for reducing total ownership cost', *Proceedings of the 2nd Joint ISPA/SCEA International Conference, San Antonio, TX.*

Ansari, S., Bell, J. and Swenson, D., 2006, 'A template for implementing target costing', *Cost Management* **20**(5), 20–27.

Ansari, S., Bell, J. and Okano, H., 2007, 'Target costing: Uncharted research territory', in C.S. Chapman, A.G. Hopwood and M.D. Shields (eds), *Handbook of Management Accounting Research*, Elsevier, Oxford, 507–530.

Ball, M., 1983, *Housing Policy and Economic Power: The Political Economy of Owner Occupation*, Methuen, London.

Ballard, G., 2008, 'The lean project delivery system: An update', *Lean Construction Journal*, Lean Construction Institute **4**, 1–19

Ballard, G. and Reiser, P., 2004, 'The St. Olaf College fieldhouse project: A case study in designing to target cost', *Proceedings of the 12th Annual Conference on Lean Construction, Denmark.*

Bartlett, E. and Howard, N., 2000, 'Informing the decision makers on the cost and value of green building', *Building Research & Information* **28**(5), 315–324.

Carmona, M. (ed), 2001, *Housing Design Quality: Through Policy, Guidance and Review*, Spon Press, New York.

Clark, K. and Fujimoto, T., 1991, *Product Development Performance – Strategy, Organization, and Management in the World Auto Industry*, Harvard Business School Press, Boston, MA.

Cooper, R., 2002, 'Target costing for new product development', *Cost Management* **16**(3), 5–12.

Cooper, R. and Yoshikawa, T., 1994, 'Inter-organizational cost management systems: The case of the Tokyo–Yokohama–Kamakura supplier chain', *International Journal of Production Economics* **37**, 51–62.

Cooper, R. and Slagmulder, R., 1997, *Target Costing and Value Engineering*, Productivity Press, Portland.

Cooper, R. and Slagmulder, R., 1999, *Supply Chain Development for the Lean Enterprise: Interorganizational Cost Management*, Productivity Press, Portland.

Cooper, R. and Slagmulder, R., 2004, 'Interorganizational cost management and relational context', *Accounting, Organizations and Society* **29**, 1–16.

Dekker, H. and Smidt, P., 2003, 'A survey of the adoption and use of target costing in Dutch firms', *International Journal of Production Economics* **84**(3), 293–305.

Deming, W.E., 1982, *Out of the Crisis*, Massachusetts Institute of Technology, Cambridge, MA.

Ellram, L.M., 1996, 'A structured method for applying purchasing cost management tools', *International Journal of Purchasing and Materials Management* **32**, 11–19.

Ellram, L.M., 2000, 'Purchasing and supply chain management's participation in the target costing process', *Journal of Supply Chain Management* **36**, 39–51.

Ellram, L.M., 2006, 'The implementation of target costing in the United States: theory versus practice', *Journal of Supply Chain Management* **42**(1), 13–26.

Everaert, P., Loosveld, S., Acker, T.V., Schollier, M. and Sarens, G., 2006, 'Characteristics of target costing: Theoretical and field study perspectives', *Qualitative Research in Accounting & Management* **3**(3), 236–263.

Feil, P., Yook, K. and Kim, I., 2004, 'Japanese target costing: A historical perspective', *International Journal of Strategic Cost Management* **2**(4), 10–19.

Flyvbjerg, B., 2006, 'Five misunderstandings about case-study research', *Qualitative Inquiry* **12**(2), 219–245.

Ford, H. and Crowther, S., 1926, *Today and Tomorrow*, Doubleday, Page & Company, New York.

Granja, A.D., Kowaltowski, D.C.C.K., Pina, S.A.M.G., Fontanini, P.S.P., Barros, L.A.F., Paoli, D., Jacomit, A.M. and Macans, R.M.R., 2009, '"A Natureza do valor desejado na habitação social" (The nature of desired value in social housing – in Portuguese)', *Ambiente Construído* **9**(2), 87–103.

Jacomit, A.M. and Granja, A.D., 2010, '"Análise crítica da aplicação do custeio-meta no desenvolvimento de empreendimentos de habitação de interesse social" (A critical analysis of the application of target costing in the development of social housing projects – in Portuguese)', *Ambiente Construído* **10**(1), 143–162.

Jacomit, A.M., Granja, A.D. and Picchi, F.A., 2008, 'Target costing research analysis: Reflections for construction industry implementation', *Proceedings of the 16th Annual Conference on Lean Construction, Manchester.*

Japan Accounting Association, 1996, 'Genka kikaku kenkyu no kadai' (The future direction of target cost management research – in Japanese), The Report of the Special Committee on Target Cost Management, Moriyama Shoten, Tokyo.

Jørgensen, B. and Emmitt, S., 2009, 'Investigating the integration of design and construction from a 'lean' perspective', *Construction Innovation: Information, Process, Management* **9**(2), 225–240.

Kato, Y., 1993, 'Target costing support systems: lessons from leading Japanese companies', *Management Accounting Research* **4**, 33–47.

Kelbaugh, D., 2007, 'Typology: an architecture of limits', in: M. Carmona and S. Tiesdell (eds), *Urban Design Reader*, Architectural Press, Burlington, 83–98.

Koskela, L., 2000, 'An exploration towards a production theory and its application to construction', Thesis (Doctor of Technology), Technical Research Centre of Finland – VTT, Helsinki, 296pp.

Kowaltowski, D.C.C.K., Silva, V.G., Pina, S.A.M.G., Labaki, L.C., Ruschel, R.C. and Moreira, D.C., 2006, 'Quality of life and sustainability issues as seen by the population of low-income housing in the region of Campinas, Brazil', *Habitat International* **30**(4), 1100–1114.

Leopold, E. and Bishop, D., 1983, 'Design philosophy and practice in speculative housebuilding: Part 1', *Construction Management and Economics* **1**, 119–144.

McNair, C.J., Polutnik, L. and Silvi, R., 2006, 'Customer-driven lean cost management', *Cost Management*, November/December **20**(6), 9–21.

Michaels, J.V. and Wood, W.P., 1989, *Design to Cost*, Wiley-Interscience, New York.

Miles, L.D., 1989, *Techniques of Value Analysis and Engineering*, 3rd edn, Lawrence D. Miles Value Foundation, Washington DC.

Monden, Y., 1995, *Target Costing and Kaizen Costing*, Productivity Press, Portland.

Monden, Y. and Sakurai, M., 1989, *Japanese Management Accounting*, Productivity Press, Cambridge, USA.

Morris, P.W.G., 1994, *The Management of Projects*, Thomas Telford, London.

Nguo, H. and Armitage, L., 2008, 'Building better suburbs: design controls in master planned communities – an Australian perspective', *Proceedings of the COBRA 2008 – The Construction and Building Research Conference of the Royal Institution of Chartered Surveyors, Dublin*.

Nicolini, D., Tomkins, C., Holti, R., Oldman, A. and Smalley, M., 2000, 'Can target costing and whole life costing be applied in the construction industry?: evidence from two case studies', *British Journal of Management* **11**(4), 303–324.

Okano, H. and Suzuki, T., 2007, 'A history of Japanese management accounting', in: C.S. Chapman, A.G. Hopwood and M.D. Shields (eds), *Handbook of Management Accounting Research*, Elsevier, Oxford, 1119–1137.

Pennanen, A. and Ballard, G., 2008, 'Determining expected cost in the target costing process', in: *Proceedings of the 16th Annual Conference on Lean Construction*, Manchester.

Robert, G.T. and Granja, A.D., 2006, 'Target and kaizen costing implementation in construction', *Proceedings of the 14th Annual Conference on Lean Construction, Santiago*.

Sakurai, M., 1989, 'Target costing and how to use it', *Journal of Cost Management* **3**(2), 39–50.

Sisodia, R., Sheth, J. and Wolfe, D.B., 2007, *Firms of Endeavor: The Pursuit of Purpose and Profit*, Wharton School Publishing, Upper Saddle River, NJ.

Sobotka, A. and Czarnigowska, A., 2007, 'Target costing in public construction projects', *Proceedings of the 9th International Conference on Modern Building Materials, Structures and Techniques, Vilnius, LA*.

Stake, R.E., 2005, 'Qualitative case studies', in: N.K. Denzin and Y.S. Lincoln (eds), *The Sage Handbook of Qualitative Research*, 3rd edn, Sage Publications, Thousand Oaks, CA.

Strauss, A.L., 1987, *Qualitative Analysis for Social Scientists*, Cambridge University Press, Cambridge.

Tanaka, M., Yoshikawa, J., Innes, J. and Mitchell, F., 1993, *Contemporary Cost Management*, Chapman and Hall, London.

Winch, G., 2002, *Managing Construction Projects: An Information Processing Approach*, 2nd edn, Blackwell Science, Oxford.

Yin, R.K., 2003, *Applications of Case Study Research*, 2nd edn, Sage Publications, Thousand Oaks, CA.

Yook, K., Kim, I. and Yoshikawa, T., 2005, 'Target costing in the construction industry. Evidence from Japan', *Construction Accounting & Taxation* **15**(3), 5–18.

Yoshikawa, T., Innes, J. and Mitchel, F., 1990, 'Cost tables: A foundation of Japanese cost management', *Journal of Cost Management for the Manufacturing Industry*, Fall, 30–36.

Yoshikawa, T., Innes, J. and Mitchell, F., 1993, *Contemporary Cost Management*, Chapman and Hall, London.

Introducing the Concept of First and Last Value to Aid Lean Design: Learning from Social Housing Projects in Chile

Christine Pasquire[1] and Jose Salvatierra-Garrido[2]

[1]School of Architecture, Design and the Built Environment, Nottingham Trent University, UK
[2]Department of Civil and Building Engineering, Loughborough University, UK

Abstract

Value for the customer through efficient production processes is a fundamental principle of lean thinking. In lean construction, value to customers is largely delivered through project planning and control activities only. Thus, it can be argued that lean construction overlooks the opportunity to address value from the early stages of a project. Aimed at improving this, lean design arose as a new approach for design management promoting customer and end user involvement from the early stage of projects. However, even here environmental and social issues are postponed over individual requirements. As a result, lean potential in general skips the opportunity to address value from a wider perspective in which the return of value from the construction industry to society is considered. This paper proposes dividing the wider understanding of the performance of the (global) built environment from the particular (local) project requirements calling the former first value and the latter last value. The theory is triangulated through observation of how a developing country (Chile) is resolving social issues through the use of the built environment. The work described develops lean design management by providing a clearer vision of value to reduce waste and aid sustainability in the built environment.

INTRODUCTION

There is a tendency in lean thinking (LT) to link the concept of value to waste reduction. This is illustrated in lean manufacturing where waste has been defined as 'specifically any human activity which absorbs resources but creates no value' (Womack and Jones, 2003, p.15). This idea also prevails in lean construction (LC) where adding value and reducing waste have been recognized as activities central to the delivery of products that satisfy customer requirements. 'Value management in Lean production is an attempt to maximise value and eliminate waste' (Bae and Kim, 2007, p.314). As a result, LC efforts have focused on improving site activities through changed project execution processes – reducing the human activity that absorbs resource and, therefore, waste. A question remains over the relationship of reducing waste in this way to actually delivering value. It is recognized that many problems – such as excessive costs, unsuccessful final results, wasteful processes, customer dissatisfaction, deficient product attributes, etc. – have their roots in inadequate management of the early design and detailing phases of a construction project (Love and Li, 2000; Brookfield *et al.*, 2004; Emmitt *et al.*, 2004; Thyssen *et al.*, 2008). LT in design and design management improves the design process through customer and end user involvement from an early stage of the project (Tilley, 2005). Additionally, lean design itself aims 'to improve the "manufacturability" of a product...concerned with generating unique value for the customer...' (Brookfield *et al.*, 2004, p.1). Consequently, LT demonstrates a close relationship between value and customer satisfaction, where the customer is commonly understood to be the person/organization/body that pays for the end product – in construction, this product is a large physical structure usually occupying or connecting land, for example, bridges can be argued to occupy air space. This understanding of customer extends into the design perspective where most efforts to deliver value have focused on satisfying the bill payer. Authors such as Emmitt *et al.* (2005) and Bertelsen and Emmitt (2005) have suggested that society is a potential customer, but they acknowledge that particular interests still predominate over social issues. The importance of developing a wider view of value for construction is being driven by the need for sustainability generally and the particular role of construction within society. There is no counter-argument to the statement that the activities of the human species *all* rely on some form of constructed facility and the more developed a country is the more this is true. It is also apparent that construction products (buildings and infrastructure) have an impact outside the immediate circle of stakeholders and that this impact has a very significant lifespan often way beyond the designed use or function. There are many examples of 'state of the art' developments that grew up to become ghettos: centres of crime, poor health and deprivation absorbing national resources, contributing little and therefore 'wasteful'. To avoid or break this wasteful decay cycle of development requires a changed view of the role of the construction sector within society – the first role of the built environment must be to contribute to society by supporting all those activities be they public or commercial, individual or group. This is challenging in several ways; the majority of society has no 'voice' in the traditional construction design process in the same way as the environment has no voice – of the sustainable triple bottom line, economic matters dominate. Also, there is little concerted effort to collect, understand and use knowledge about the lifecycles of development to influence the design and management of the built environment as a whole even where planning and regulation exist. These provide checks and balances to aspects of development but joined up urban and rural planning governing more than blocks of land use is limited in the developed world – all major decisions were made long ago. The situation is different in the developing world where opportunities exist to design a built environment that supports society and impacts directly on the national economy. Here the relationship between these is more visible, and cause and effect over time can be measured.

RESEARCH METHODOLOGY

The research described in this article comprised a desktop literature review and a case study comprising two social housing projects (SHP) in Santiago, Chile. These projects were initiated in response to a new social housing policy that intended to insert the homeless population into local society. This policy demanded more than simply providing accommodation of a stated quality but considered social impact by requiring the developments to contribute to the society into which they would be integrated. The literature review has sought to explain how this policy fits into the accepted understanding of value in construction. The attributes of the case study projects were identified through examination of the SHP documentation. Interviews with a variety of stakeholders were used to triangulate the findings from the literature and illustrate the inclusion of social value into the development of the built environment.

REVIEW OF VALUE IN CONSTRUCTION LITERATURE

The concept of value is widely discussed and, according to Leszinski and Marn (1997), 'one of the most overused and misused concepts of social sciences in general and in management literature in particular'. LC authors agree that it is difficult to discuss and define (Bertelsen and Koskela, 2004, p.6; Emmitt *et al.*, 2004, p.3). Previous research (Salvatierra-Garrido *et al.*, 2009; Salvatierra-Garrido *et al.*, 2010) aiming to better understand the concept of value from a construction perspective revealed two key findings:

- It is necessary to differentiate between value and *values*. These concepts are commonly misunderstood as one being the plural of the other. Value can be understood as what a customer places upon a product (Wandahl, 2005), whereas *values* are innately subjective; they make human guidelines (ethical and ideological) evident, representing the value system of an individual, organization or society as a whole (Amis *et al.*, 2002; Thomson *et al.*, 2003a; Kelly *et al.*, 2004; Thomson and Austin, 2006; Glew, 2009, p.675).
- Value has been mainly understood as an objective concept where numerous parameters such as cost, quality, function, etc. have been associated to value delivery (Green, 1997; Thomson *et al.*, 2003a,b). Some subjective parameters such as design have been linked to this concept, and architects are universally considered to be carriers of the customer value perspective (Erikshammar *et al.*, 2010). To address value in the construction sector in a better way, the full range of its diverse features should be considered: objectivity, subjectivity, relativity, dynamism, context dependence (Wandahl, 2005).

VALUE FROM AN LC PERSPECTIVE

From the beginning of LT in the manufacturing sector, the word value has been introduced as a concept to analyse the performance of production processes. Thus, in the book *Lean Thinking* the concept of value was defined as 'a capability provided to a customer at the right time and at an appropriate price, as defined in each case by the customer' (Womack and Jones, 1996, p.311). In addition, it has been recognized that LT practices 'must start with a conscious attempt to precisely define value in terms of specific products with specific capabilities offered at specific price through a dialogue with specific customers' (Womack and Jones, 2003, p.19). When LT was adapted to the environment of the construction sector, this conception of value as materials, parts or products was broadly criticized (Koskela, 2000, 2004). However, the LC perspective of value is still strongly influenced by the lean production theory. In his deep investigation, Koskela (2000) distinguished three theoretical models to show construction as a production process: transformation, flow and value generation (TFV). As a consequence TFV is presented as a theory of production for the construction sector. To date, LC experience has demonstrated that the concepts of transformation and flow have been widely applied across the supply chain activities: transformation meaning where input is transformed into output along on-site activities, and flow in the sense of improvement in control and organization and planning activities. However, value generation has been less explored focused only on the fulfilment of a particular customer's requirements, through control and planning of construction activities. Therefore, value has been associated with objective parameters, such as waste reduction, cost, quality, time, function, etc. of activities performed at project level intended only to satisfy customer requirements.

VALUE FROM A LEAN DESIGN MANAGEMENT PERSPECTIVE

When LT was translated as a management process for design, activities were targeted mainly towards improving the manufacturability of products. Thus, the early stages of a project focused on coordination and flow of information, and on the development of designs capable of satisfying technological, functional and operational requirements (Brookfield *et al.*, 2004). In addition, the earlier phases of lean design should generate options to enhance client values and go beyond the mundane to create 'real value for client and building users alike' (Emmitt *et al.*, 2004, p.2). To avoid problems arising in the construction stage, lean design management has also been linked to the improvement of communication and the decision-making process. In this way, interactive workshops with participation of stakeholders become a central activity to create consensus in the differing value judgements. These activities are understood as an opportunity to reduce future uncertainty and minimize waste (Emmitt *et al.*, 2004). The interaction of different stakeholders allows consideration of the notion of an expanded customer/client. Different bodies such as owners, constructors, subcontractors, end users, authorities, etc. contribute with different value judgements from the early stage of projects. This expanded customer notion provides an opportunity to consider society as a potential customer. In this way, some lean design authors have emphasized the importance of considering society and social aspects:

- 'The client often represents a lot of different stakeholders (the users, the investor, the owner etc.), and furthermore when we build we affect our neighbour and the surroundings (city/landscape etc.) and they all have a different set of values and interests in the project' (Emmitt *et al.*, 2005, p.59).
- 'Looking closer at the nature of the client, one may identify behind the façade an organisation, which during the project execution should represent the interests of three distinct client groups – the owners, the users and society. These three groups of interest each value different things at different time in the life of the building' (Bertelsen and Emmitt, 2005, p.74).

So lean design management clearly includes ensuring timely interaction of the appropriate parties. Moreover, other authors in the field of LC have underlined the potential of including society from a lean perspective (Koskela, 2000; Dugnas and Oterhals, 2008; Höök, 2006; Alves and Tsao, 2007; Forsberg and Saukkoriipi, 2007; Koskela *et al.*, 2008). Ballard (2008) believes there is a growing tendency to consider sustainability in construction design and that whole life issues are becoming more important. Despite the strength of these calls, general practice continues to postpone social issues for particular interests. Ballard (2008) adds that using a lean project delivery system to manage definition and design of projects will deliver value within the constraints. In addition, the supply chain commonly understands the customer as the *one who pays* for satisfying particular requirements. As a result, social issues continue to be included only if they are important from the bill payer's perspective – one would expect this to be strongest in public sector construction. To address this, the needs of society should be integrated into the agenda of the bill payer in a way that provides a voice to the voiceless. It is proposed that value is visualized through the introduction of two new concepts, *first value* and *last value*, in order to elevate social issues in design management.

FIRST AND LAST VALUE CONCEPTS

LT requires an end-to-end view of the system under consideration. The construction sector is compartmentalized – there are connections but little integration between design, build/construct and repair/maintain. To make matters worse, there is a clear divide between these three activities and the planning of the built environment and the use of the built environment. This separation is entirely responsible for the absence of social issues on the bill payer's agenda – what has built environment planning and use to do with the construction customer and supply chain other than to constrain choice? As a result there is a general failure to recognize the impact of the construction sector output as a whole on society (and vice versa) beyond that required by planning legislation. The concepts of first and last value intend to (re)connect society with the construction process and evidence a return of value to society as a result. These concepts recognize value as an *oscillating* phenomenon. That means, value delivered for a particular building project impacts society as a whole and consequently the judgements of value for future building projects, and so on and so forth (Salvatierra-Garrido *et al.*, 2010). It has been proposed that architects are the carriers of customer value (Erikshammar *et al.*, 2010) – the details of first and last value, however, are too big for one person to know and own. Responsibility for first and last value belongs with the design manager who needs to ensure that parties with the appropriate knowledge are engaged in the process. A *lean* design manager should have an end-to-end view clearly recognizing when and how to engage with planners and/or politicians as well as special interest groups, end users and members of the supply chain.

FIRST VALUE

This concept considers value delivery in a wide sense, where the impact of the construction industry over society is considered. This classification of value proposes an expansion of common customer-focused perspectives to cover broader aspects – buildings and infrastructure as social assets, development as it contributes to cures for both environmental and social issues (e.g. climate change, sustainability, pollution, reducing crime, increasing educational levels, social network insertion, socialization, etc.). Here the widest sense of societal benefit includes sustaining the environment and the economy for human survival and development. First value therefore expresses societal goals, and is high level, politically led and somewhat intangible.

LAST VALUE

This concept considers value delivery at production level, linked to transformation activities or what needs to be done (the work). Therefore, this concept is easily observable through current LC practices, where value has been mainly associated to market and utility value. Thus, *last value* can be commonly understood by the supply chain as what customers will pay for products or services (what customers give to achieve an unsatisfied need).

The two concepts are not discrete but rather exist at either end of a scale. Not every construction project needs to fulfil the aspirations of first value – indeed few will, but all projects must fulfil the ends defined under last value. The intention of this work is to enable the underlying performance of the construction sector to move towards a more considered contribution to society (i.e. to the left in Table 1) by conceptualizing the 'big picture' or social system. Continual improvement is a fundamental lean principle so first value is intended to form an aspiration rather than a tangible set of criteria: in the way that perfection is a final lean principle (Womack and Jones, 2003). This scale of first to last value provides context to the existing thinking on value found in the literature on lean design presented in Table 1.

Product values are commonly understood as 'derived from Vitruvian values (firmness, commodity and delight), combined with harmony with the surrounding, environmental issues and buildability' (Emmitt *et al.*, 2005, p.59). According to Gann *et al.* (2003), function, build quality and impact can be the modern interpretation of Vitruvian values. Based on design quality indicators, the modern terms

TABLE 1 Visualization of first and last values compared with current design value

AUTHOR/S	DESIGN VALUE[a] CLASSIFICATION	SCALE FOR ILLUSTRATION – NOT MEASURING QUALITY OR QUANTITY							
		FIRST VALUE		⟺			LAST VALUE		
Gann *et al.* (2003)	Function						X	X	X
Vitruvian values	Build quality							X	X
represented by design	Impact			←	X	X	X	X	X
quality indicators									
Emmitt *et al.*	External value								
(2005, p.59)	• Product						X	X	X
Vitruvian values	• Process:								
encompassed within	hard							X	X
product value	soft						X	X	X
	values	←	X	X	X	X	X	X	X
	Internal value								X
Ballard (2008)	Lean project delivery			←	X	X	X	X	X
Ends, means and constraints	system								

←, this aspect already demonstrates the potential to deliver first value.

X, degree to which value fulfilment goes beyond meeting bill payer requirement.

[a] Description of the design values.

associated with Vitruvian values can be defined as follows:

- Function: encompasses aspects of its use, access and space;
- Build quality: encompasses aspects of its performance, engineering system and construction; and
- Impact: encompasses aspects of its contribution to form and materials, the internal environment, urban and social integration, identity and character' (Gann *et al.*, 2003).

Emmitt *et al.* (2005, p.59) go further, dividing value into:

- external value, which is the client/customer value, and the value that the project should end up with and the delivery team focusing on achieving;
- internal value, by and between the participants of the delivery team.

In addition, external value is sub-classified as:

1 product value demonstrated through the final physical product and related to Virtuvian values; and
2 process value, which is about giving out customers the best experience during design and construction of projects. It comprises:
   - 'soft values' such as work ethics, communication, conflict solving, etc. between the client and the delivery team;
   - 'hard values' such as the delivery team's ability to keep to agreed time limits, cost estimates, quality of the product, workers' safety, etc.;
   - values that come from the actual design and construction process. As an example of this kind of value, renovation works in a kindergarten could be used to teach children about safety, creativity, etc. and thus generate process value that might not have been evident when the project started out. Learning from participating in the process is another value in this category.

Ballard (2008) places value as an outcome to the customer derived from the purpose of the development. As such this aligns with the definition proposed by Gann *et al.* (2003), but Ballard expresses this as a combination of end, means and constraints. These classifications provide the current visualization of value from a construction industry perspective. The authors cited above also consider a wider set of customers including owner, user and society – but society is commonly represented by authorities as an external constraint not an integral part of the design creation.

To investigate whether first and last value have any potential application, two SHP in Chile were investigated.

SHP IN CHILE

As a developing country, Chile provides a fresh platform for designing a built environment that can deliver a measurable contribution to society. Public and private organizations are working under the New Chilean Housing Policy (NChHP) of 2007 to meet housing requirements and to build cities and neighbourhoods that will improve the quality of life, unlike previous efforts that were focused merely on reducing the housing deficit. The NChHP not only intends to reduce the number of people living in precarious conditions, but is also concerned about the dignity of those people (Poblete, 2007). This research investigated those features and attributes of the SHP that indicated a wider view of value and considered the return of value from construction projects to society as a whole.

A review of the government document *Fondo Solidario de Vivienda-Manual para dirigentes y familias* (*Housing Support Fund – Handbook for Leaders and Families*) (www.minvu.cl) revealed three central aspects:

1 *Principles of the NChHP*
   a. reduction of the deficit, focused on people living in poor conditions;
   b. guarantee of quality, improving the standard, the design and construction processes; and
   c. promotion of social integration, procuring solutions located in neighbourhoods and cities.
2 *Activities throughout the lifecycle of projects according to the NChHP:* There are three main objectives: (1) delivery of the house, (2) improvement of housing standards and (3) insertion of families

into society. These can be expressed in two themes:

First, the house and its attributes:

a. selection of building site;
b. definition of project budget;
c. definition of project features;
d. project's design;
e. post-sale services;
f. revision and qualification of projects (technical, economic and legal);
g. planning and controlling construction site activities.

Second, social insertion:

i. detection of social problems: delinquency, drug abuse, unemployment, etc.;
ii. definition of social insertion plan (SIP);
iii. family professional training;
iv. insertion of people in social networks (schools, hospitals, transportation, police station, etc.).

3 *Stakeholders' responsibilities according to the NChHP:* According to the NChHP, six groups of stakeholders are involved throughout the lifecycle of projects. Their responsibilities are defined as follows:

A. Families (householder):
   - transmit requirements;
   - complete savings requirement;
   - participate in activities of the social insertion plan (SIP).
B. EGIS (private organization dedicated to advising and guiding families on technical and social requirements in order to obtain satisfactory final solutions):
   - represent families' requirements;
   - design the project or supervise its design by the architectural team;
   - contract building company services;
   - coordinate project execution;
   - determine completion of quality requirements;
   - collect documentation to obtain monetary assistance from the government (subsidy);
   - elaborate technical and social project documentation to submit them to SERVIU;
   - design and carry out activities related to the SIP;
   - manage final reception of projects;
   - carry out legal proceedings and register new house in order to authenticate these processes.
C. Architectural team:
   - design the project.
D. Building company
   - build the project.
E. *SERVIU*: Government department responsible for housing programmes and for reviewing SHP. It also pays for the EGIS's work and distributes subsidies for building new SHP:
   - review, qualify and approve projects;
   - provide funds to pay for technical and social assistance;
   - distribute monetary assistance from the government (subsidy);
   - carry out or contract technical inspection during construction process;
   - town council (the local government);
   - collect information about families at social risk;
   - provide information on building land (cost and availability);
   - allow construction of the projects;
   - approve final reception of the projects.

CASE STUDY PROJECTS: STAKEHOLDERS' ACTIVITIES

To complement the documentary information about the NChHP, two SHP in Chile were investigated. Neither project was built explicitly using LT:

1 'Campamento: Vista Hermosa – Un Sueño por Cumplir': Located in the Town Council 'Lo Espejo', the project to house 300 families began in 2000. The area was part of the biggest shantytown of the Metropolitan Region of Santiago of Chile. In December 2007, the first 30 families grouped under the committee named 'Sueño por Cumplir' received their houses.
2 'Campamento: Lo Boza – Condominio Antumalal': Located in Town Council 'Renca', the project was created to house 170 families coming from four

shantytowns of this district. In June 2008, these families received their houses.

In order to capture stakeholder experience, eight interviews were carried out with senior members from stakeholder organizations B–F, using two semi-structured survey instruments. Each instrument comprised open-ended questions designed to gain an insight into the context for the consideration of value.

INTERVIEW SURVEY 1

The interview comprised two open-ended questions to identify the particular and general stakeholder experience:

1 In relation to families, which needs are you trying to meet?
2 In relation to society, which needs are you trying to meet?

The list below summarizes the data collected:

- housing provision;
- provide basic facilities, toilet, kitchen, etc.;
- provide basic infrastructure for further extensions;
- improve families' standard of life (quality of life);
- reduce number of non-paying lodgers;
- protection of vulnerable groups (social vulnerability/risk);
- provide public safety;
- minimize urban poverty;
- avoid stigmatization associated with neighbourhood's location;
- avoid residential and social segregation;
- improve social environment (aesthetic).

INTERVIEW SURVEY 2

Intending to gain an insight into the activities carried out at the early stage of projects, stakeholders were asked:

- Based on the job performed, detail the activities carried out in the following stages of the project: new building required, feasibility, scheme design and detail design.

  The activities identified were:

  - New building requirement:
    - coordinate activities;
    - determine demand;
    - provide for participative workshops (e.g. housing design, neighbourhood, equipment, etc.);
    - address families' expectations;
    - assess land site alternatives.
  - Feasibility:
    - evaluate project costs and feasibility aspects, technical and legal studies in relation to the site;
    - determine the exact number of families to be included;
    - define housing size, model and urban sketches;
    - determine economic feasibility (design).
  - Scheme design:
    - workshops (housing design, neighbourhood, equipment);
    - management of families' expectations;
    - elaboration of models to scale and 3D models (design based on the use of the house);
    - elaboration of speciality projects (sanitation, paving and electricity).
  - Detail design:
    - adjustment of budget design;
    - approval of final design;
    - certification of final solution (technical and legal requirements);
    - elaboration of housing plan (technical drawings);
    - procurement of edification permissions.

FIRST AND LAST VALUE IN SHP IN CHILE

NChHP's principles give the opportunity to visualize value from a wider perspective. The first two principles (reduction of deficit and quality) are closely related to last value delivery through the provision of a house and its attributes. However, the third principle clearly shows value delivery to the entire society (towards first value). Comparison of the first and last value concept with data collected describes how these can be seen within the case study projects.

LAST VALUE

The first two principles of the NChHP give the opportunity to address value delivery at project level (value associated with product delivery and product attributes – the house). The Chilean experience demonstrates that *last value* can be visualized as:

- provision of the house;
- management of monetary, human and natural resources;
- improvement of housing attributes (location, quality, materials, function, etc.);
- management of product attributes in function to available resources (cost and time);
- planning and control of activities (design and construction);
- stakeholders' involvement from the early stage of projects (provide an interpretation of families' vision on housing expectations).

FIRST VALUE

The third principle of the NChHP includes value from a wider perspective. Consequently, the analysis of activities and responsibilities associated with different stakeholders is clearly oriented towards the cure of social issues such as delinquency, drug abuse, unemployment, etc. There is a close relationship between SHP and the welfare of the entire population. In the case studies, the construction industry delivers value to the entire society and the phenomenon where value is understood as an *oscillating* concept applies and the decisions made will inform future projects. In this way, the Chilean experience demonstrated a move towards *first value*, which can be visualized as:

- insertion of low-income people into society;
- less stigmatization;
- less social vulnerability and risk;
- improvement of public safety;
- improvement of social environment (aesthetic), etc.

CONCLUSIONS

LC theory indicates that most efforts have been made to deliver value from a customer-focused perspective, where project control and planning are central activities. Recognition of the need for a softer and wider definition exists and this should include sustainability and whole life issues – the needs of society are included within these. The first and last value approach adds several aspects to the theory so far:

- That society – considered as human survival and development – encompasses environmental and economic sustainability as society is dependent on these.
- That value so far is seen as a self-contained project attribute/set of attributes that can be defined; allowing value to be visualized along a scale permits wider goals and responsibilities to be recognized and a strategy formulated.
- That visualizing value on a scale enables a project design to be questioned 'how are we doing on the first value scale?' as design progresses. This pushes design to the left (in Table 1) by requiring some view of first value for each project.
- That visualizing value on a scale (last to first) moves project design towards increasing value – aligning with the lean vision of more for less.
- That recognizing the oscillating nature of value with the strategy of social goals influencing projects which feed benefit back to society and so on will contribute to a lean end-end view for the management of the built environment and urban and rural planning.

The investigation of SHP in Chile demonstrated the possibility and desirability of developing a construction strategy that delivered a return of value to society. These projects aimed to improve healthcare, reduce crime, increase employment, integrate low-income people into social networks, etc. and the design was managed accordingly (although not expressly lean) and provides tangible evidence that social issues can drive development.

Finally, this article forms part of research whose objective is to visualize value from a wider perspective with the objective of returning value from projects to society as a whole. However, there are questions that need further investigation. For example:

- How does *last value* translate into *first value*?
- Can companies be rewarded other than by payment of money for delivering *first value* to society as a whole?

- If not, how can payment for *first value* be made to better reflect *last value*?

ACKNOWLEDGEMENTS

The research reported in this article is part of an ongoing PhD of Loughborough University – Department of Building and Civil Engineering. The authors would like to acknowledge the support for this research from all the interviewees involved in social housing projects through the following institutions: Ministry (Sub-Department of Housing Operations – SERVIU Metropolitano), Lo Espejo-Town Council, Architectural company (Elemental), Building company (Simonetti), EGIS (UTpCH-Definitive Housing Area UTpCH), Renca-Town Council, Building company (SIESCON Constructora), EGIS (JUNDEP).

REFERENCES

Alves, T. and Tsao, C., 2007, 'Lean construction – 2000–2006', *Lean Construction Journal* **3**(1), 46–70.

Amis, J., Slack, T. and Hinings, C., 2002, 'Values and organizational change', *Journal of Applied Behavioural Science* **38**(4), 436–465.

Bae, J. and Kim, Y., 2007, 'Sustainable value on construction project and application of lean construction methods', *Proceedings of the 15th Annual Conference on Lean Construction*, IGLC, *East Lancing, MI, USA*, 312–321.

Ballard, G., 2008, 'Lean project delivery system – an update', *Lean Construction Journal* 1–19.

Bertelsen, S. and Emmitt, S., 2005, 'The client as a complex system', *Proceedings of the 13th Annual Conference on Lean Construction*, IGLC, *Elsinore, Australia*, 73–79.

Bertelsen, S. and Koskela, L., 2004, 'Construction beyond lean: a new understanding of construction management', *Proceedings of the 12th Annual Conference on Lean Construction*, IGLC, *Elsinore, Denmark*, 1–12.

Brookfield, B., Emmitt, S., Hill, R. and Scaysbrook, S., 2004, 'The architectural technologist's role in lean thinking design with lean construction', *Proceedings of the 12th Annual Conference on Lean Construction*, IGLC, *Elsinore, Denmark*, 1–13.

Dugnas, K. and Oterhals, O., 2008, 'State-of-the art shipbuilding: towards unique and integrated Lean production system', *Proceedings of the 16th Annual Conference on Lean Construction*, IGLC, *Manchester, UK*, 321–331.

Emmitt, S., Sander, D. and Christoffersen, A.K., 2004, 'Implementing value through lean design management', *Proceedings of the 12th Annual Conference on Lean Construction*, IGLC, *Elsinore, Denmark*, 1–13.

Emmitt, S., Sander, D. and Christoffersen, A.K., 2005, 'The value universe: defining a value based approach to lean construction', *Proceedings of the 13th Annual Conference on Lean Construction*, IGLC, *Sydney, Australia*, 57–64.

Erikshammar, J., Björnfot, A. and Gardelli, V., 2010, 'The ambiguity of value', *Proceedings of the 18th Annual Conference on Lean Construction*, IGLC, *Haifa, Israel*, 42–51.

Forsberg, A. and Saukkoriipi, L., 2007, 'Measurement of waste and productivity in relation to lean thinking', *Proceedings of the 15th Annual Conference on Lean Construction*, IGLC, *Michigan, USA*, 67–76.

Gann, D.M., Salter Ammon, J. and Whyte Jennifer, K., 2003, 'Design quality indicator as a tool for thinking', *Building Research and Information* **31**(5), 318–333.

Glew, D., 2009, 'Personal values and performance in teams: an individual and team-level analysis', *Small Group Research Journal* **40**(6), 670–693.

Green, S.D., 1997, 'New directions in value management', *Proceedings of the Hong Kong Institute of Value Management International Conference: Effective Change through Value Management, Hong Kong*, 1–8.

Höök, M., 2006, 'Customer value in lean prefabrication of housing considering both construction and manufacturing', *Proceedings of the 14th Annual Conference on Lean Construction*, IGLC, *Santiago, Chile*, 583–594.

Kelly, J., Male, S. and Graham, D., 2004, *Value Management of Construction Projects*, Blackwell Science, Oxford.

Koskela, L., 2000, 'An exploration towards a production theory and its application to construction', PhD thesis, Technical Research Centre of Finland-VTT, Helsinki.

Koskela, L., 2004, 'Moving on - beyond lean thinking', *Lean Construction Journal* **1**(1), 24–37.

Koskela, L., Siriwardena, M. and Rooke, J., 2008, 'Through-life management of built facilities: towards a framework for analysis', *Proceedings of the 16th Annual Conference on Lean Construction*, IGLC, *Manchester, UK*, 61–71

Leszinski, Marn, 1997.

Love, P.E.D. and Li, H., 2000, 'Quantifying the causes and costs of rework in construction', *Construction Management and Economics* **18**, 479–490.

Poblete, 2007, P-MINVU Primeras viviendas de nueva política habitacional en Región Metropolitana [available at www.minvu.cl/opensite_det_20071207160817.aspx] (viewed 30 April 2008).

Salvatierra-Garrido, J., Pasquire, Ch. and Thorpe, T., 2009, 'Value in construction form lean thinking perspective: current state and future development', *Proceedings of the 17th Annual Conference on Lean Construction*, IGLC, *Taipei, Taiwan*, 127–136.

Salvatierra-Garrido, J., Pasquire, Ch. and Thorpe, T., 2010, 'Critical review of the concept of value in lean construction theory', *Proceedings of the 18th Annual Conference on Lean Construction*, IGLC, *Haifa, Israel*, 33–41.

Thomson, D. and Austin, S., 2006, 'Using VALiD to understand value from the stakeholder perspective', *Proceedings of 46th SAVE International Annual Conference, Savannah, GA*, 1–11.

Thomson, D.S., Austin, S.A., Devine-Wright, H. and Mills, G.R., 2003a, 'Managing value and quality in design', *Building Research & Information Journal* **31**(5), 334–345.

Thomson, D.S., Austin, S.A., Mills, G.R. and Devine-Wright, H., 2003b, 'Addressing the subjective view of value delivery', *COBRA 2003*, University of Wolverhampton, The RICS Foundation.

Thyssen, M.H., Emmitt, S., Bonke, S. and Christoffersen, A.K., 2008, 'The Toyota product development system applied to design management workshop model', *Proceedings of the 16th Annual Conference on Lean Construction*, IGLC, *Manchester, UK*, 507–518.

Tilley, P.A., 2005, 'Lean design management – a new paradigm for managing the design and documentation process to improve quality?', *Proceedings of the 13th Annual Conference on Lean Construction*, IGLC, *Sydney, Australia*, 283–295.

Wandahl, S. 2005, 'Value in building', PhD thesis, Aalborg University, Denmark.

Womack, J. and Jones, D., 1996, *Lean Thinking: Banish Waste and Create Wealth in Your Corporation*, Simon & Schuster, New York.

Womack, J. and Jones, D., 2003, *Lean Thinking: Banish Waste and Create Wealth in Your Corporation*, 2nd edn, Simon & Schuster, New York.

Integration of Lean Design and Design Management and its Influence on the Development of a Multidisciplinary Design Management Model for Remote Site Projects

Linda Kestle[1], Regan Potangaroa[2] and Bryan Storey[3]

[1]Faculty of Technology and Built Environment, Unitec Institute of Technology, New Zealand
[2]Faculty of Creative Industries and Business, Unitec Institute of Technology, New Zealand
[3]Gateway Antarctica - Centre For Antarctic Studies and Research, University of Canterbury, New Zealand

Abstract

A significant body of literature addresses the application of lean thinking to improving the interface between detailed design and construction production, yet none specifically focuses on remote site projects. These projects range from tourism and scientific investigation to resource exploration developments. The challenges confronting the project teams tend to fit within a *sociologically oriented world* where designers respond to functional, aesthetic and environmental concerns, or within a *production oriented world*, where strategic decisions made during the early stages of a project impact markedly upon construction, logistics and sustainability. The research aimed to establish how the integration of lean design and design management thinking influenced the development of a conceptual design management model for remote site projects, and the level of rigour achieved 'in the field' when tested. A postdoctoral reflective review is made of a selection of the reviewed literature, the methodology/process undertaken in the model development and testing stages, and the findings from two case studies used when testing the developed model. Semi-structured interviews were conducted with selected 'working-in-the-field' participants across a diverse range of remote site projects; the model was found to be robust and portable.

INTRODUCTION

Projects in remote locations such as the construction of dams, canals, oil rigs and the rebuilding work associated with disaster relief have been undertaken for a number of years now, with disaster relief and reconstruction projects markedly increasing in the past 5 years or so in Africa, Malaysia, Pakistan and China. These projects appear to have several features in common:

- Multi-stakeholders sometimes form a management, operational and collaborative consortium (a) for the duration of the project.
- The work is often conducted under extreme weather/climatic conditions, tight timeframes and limited site accessibility.
- The processes associated with the environmental context and potential impacts appear to frequently involve significant

and protracted protocols, and regulatory frameworks.
- Political agendas (and therefore influence) also appear to be evident on many if not all of these projects.

Increasing numbers of remotely located and often environmentally sensitive sites are becoming the focus of new or post-disaster development work involving potential investors/entrepreneurs/stakeholders or government and non-government agencies. In addition, clients, stakeholders, designers and construction industry representatives involved on these frequently sensitive remote sites now have an increasing 'duty of care', in a global sense, to these pristine environments (Kestle, 2009).

The specifics associated with researching into design management and remote sites rendered no previously documented empirical examples, theoretical models or theories related to this specific topic, from an integrated design and management perspective. Theory-building and model-testing was therefore seen to be required in this field/discipline of managing projects on remote, environmentally sensitive and often hostile sites (Kestle, 2009). A body of literature now addresses the application of lean thinking to improving the interface between detailed design and construction production, and this is seen as being conducive to the complex process involved in the design management of remote site projects.

AIM

The aim of this work was to reflect on how the integration of lean design management and design management thinking informed and influenced a conceptual design management model developed for internationally collaborative remote site projects, and the level of rigour achieved 'in the field', when tested.

REFLECTIONS ON A SELECTION OF THE LITERATURE REVIEWED

The literature reviewed over the full course of the conceptual management model's development and testing within the context of the doctoral research included the following topics/disciplines:

- design management current practice and issues;
- lean design management, lean thinking and production;
- humanitarian aid management;
- project management and value management;
- environmental sustainability and sustainable development.

Only a selection of the reviewed and analysed literature is included here for clarity and to address the lean and design management integration focus of this article. The complete literature review and analysis is given in the doctoral thesis (Kestle, 2009).

DESIGN MANAGEMENT

Design management is fundamentally concerned with value generation for the customer/client, which involves integration of specialist knowledge, and the timing of key decisions. This is achieved by means of an integrated team approach to the way in which the project is designed, constructed, implemented and managed. In the case of remote site projects, this involves a necessary interface between two quite different process perspectives: the iterative creative process, which tends to be a *sociologically oriented world*; and a perspective that involves a *production-oriented world* which addresses strategic decisions during the early design and planning stages. How these could impact upon construction logistics, and the sustainable development of the project site, is also considered.

Understanding what constitutes value is a difficult process that is never simple or straightforward. Instead, design management is a complex social situation, as value can be a socially constructed phenomenon and decision-making to that end can be inherently unpredictable (Kestle and London, 2002).

A shared understanding by all of the stakeholders of what is valued in the project has an impact on how critical decisions are made on design issues (Tombesi, 1997). The integration of a team with knowledge that can contribute to the design, construction and management is critical to developing and achieving value on projects for the client and stakeholders. Poor integration of specialist user and producer stakeholder knowledge can have far-reaching consequences, such as inappropriate synthesis of the needs analysis leading to low value generation for the client and the end-users (Kestle, 2009).

On reflection, this analysis is found to still be current as supported by the fact that there are ongoing investigations into the concept of value and value generation, which continues to be elusive and significant for design management and lean researchers and practitioners. This was demonstrated recently, for example, by Salvatierra-Garrido *et al.* (2010), who noted that the concept of value varies across time, is context dependent, is relative/comparative and very subjective, and tends to be restricted to just achieving value for end-users and clients rather than society as a whole. Salvatierra-Garrido *et al.* (2010) continued, stating that:

> Authors such as Ballard (2006) have been working in the adoption of a model of project definition with a value generation perspective. This new model gives relevant importance to the stakeholders' perspective of value. Development of the 'Target Value Design' model by Ballard is ongoing, . . . and the detailed value discussion has yet to be published. Stakeholders' involvement is considered a key element in generating Value, therefore, it is important to underline the contribution of Emmitt *et al.* (2004), where the concept of Value is divided into: External value, which is the client/customer value, the value that the project should end up with and the delivery team focus on achieving, and Internal value by and between the delivery team.

The holistic approach to lean design management explored by researchers such as Green (1994), Huovila and Koskela (1998), Garnett (1999) and London (2002) identified additional significant design management factors. These researchers referred to the importance of sustainable development and the ways to achieve it. They believed that while traditional design and construction focuses on cost, performance and quality objectives, sustainable design and construction, by comparison, focuses on value generation, minimization of resource depletion, minimization of environmental degradation and the importance of timely information flow management (Kestle, 2009). In the discussion regarding differing models that can provide a managerial framework for construction projects, Emmitt *et al.* (2004) referred to the 'value-based design model', the key features of which strongly resonate with the holistic aspects of the practice and discipline of design management.

LEAN THINKING AND LEAN PRODUCTION

The concepts and implementation of the key lean design principles and lean production are 'value stream', 'process integration', 'workflow' and 'waste minimization' (Womack and Jones, 1996). These have been applied and specifically interpreted within the construction industry worldwide for more than four decades. The previous design and value management literature in Kestle (2009) and Howell and Ballard (1998) discussed the goals of lean thinking, and the early misconceptions and resistance among construction industry personnel regarding lean thinking as a useful management tool, with lean thinking often seen as being simply a manufacturing technique.

Howell and Ballard (1998) 'believed that the goals of lean thinking did in fact describe the management of dynamic projects' and went on to state that 'Lean is a value seeking process that maximises value and continually redefines perfection' and that 'the goals of lean thinking redefine performance against three dimensions of perfection:

- a uniquely custom product;
- delivered instantly; and
- nothing in stores'.

This in their view maximizes value and minimizes waste. This third dimension of perfection certainly had relevance for the Antarctic sites in particular. They further suggested that 'Lean thinking forces attention on how value is generated rather than how any one activity is managed' (Howell and Ballard, 1998), and that 'lean production presents a very different model – where production is managed so that actions are aligned to produce unique value for the customer'. This, they said, was basically achieved by ensuring that value to the customers, throughput, management, information and materials required to completion were the primary objectives.

Conversely, Howell and Ballard (1998) suggested that project management was essentially a combination of activities, stating that:

> lean thinking views the entire project in production system terms, that is one large operation ... and implementing lean means adopting a 'project-as-production-system' approach ... defining the objective in customer terms and decentralizing management.

The lean thinking and lean production tenets identified by Womack and Jones (1996) included 'specifying value by product', thereby essentially placing the specification of value ahead of design; and 'identifying the value stream', which influenced when and how decisions are made. Mapping of the 'value stream' identifies which actions determine when the next operation can start, according to Howell and Ballard (1998). This could be regarded as having similarities to critical path analysis for task management planning.

Lean thinking aims to translate local-level improvements into better results for the whole project, by reviewing every stage as a part of the whole and identifying their impacts on the outcomes. These discussions and debates have continued among researchers in resultant publications at the annual International Group for Lean Construction (IGLC) conferences or in the *Lean Construction Journal*. For example, Koskela (2004) suggested that there was a need to move on from lean thinking 'toward a more generic theory of Lean production to provide a more solid foundation for design, operation and improvement of production systems'. Koskela (2004) then provided a critical assessment of the premise that lean thinking was an underpinning theory for lean production, concluding that 'lean thinking cannot be viewed as a mature and valid theory of (lean) production'. A second recent example of continuing challenges to the application of lean thinking in the construction context was found in the work by Jorgensen and Emmitt (2009), where 'a multiple case-strategy approach was used to explore the practical application of lean approaches to design and construction integration in an organisational setting'.

LEAN DESIGN MANAGEMENT

In earlier lean design management literature, 'value' was often seen as the sole focus of research investigations and found to be an important part of design management. However, other models were explored by Formoso *et al.* (1998), who included 'conversion' and 'flow' in their investigations, and by Ballard and Koskela (1998) and Freire and Alarcon (2000), who concluded that three distinct models – conversion, flow and value generation – constituted the process of lean design. An analysis of the application of some of the lean construction principles to design management from the point of view of design as conversion, flow and value generation was made in an article by Tzortzopoulos-Fazenda and Formoso (1999). The findings of the two Brazilian case studies in their article identified some gaps in the knowledge of the application of theory in design and in particular the value generation view of design concepts and principles.

The lean design management literature primarily focuses on the sequential production approach and processes. Researchers Garnett (1999) and Huovila and Koskela (1998) adopted a more sociological approach to lean design, where the principle of 'flow' is relevant from a sociological and environmental viewpoint, as it tends to be focused on a more holistic approach for theoretical and project development work. This type of approach resonates with more remote sites, which are frequently environmentally sensitive. Ballard and Koskela (1998) and Koskela *et al.* (2002) suggested that the way forward was to simultaneously view design management as conversion, flow and value generation, rather than the more traditional view of engineering as a conversion. The 'conversion view' is basically about 'getting the task done' using practices such as 'work breakdown structure' and the 'critical path method', where 'flow and generation' and even the client are not considered in the programming, or outcomes. Of specific interest for remote site design management was the comment by Ballard and Koskela (1998) and Koskela *et al.* (2002) that 'what is needed is a management philosophy and tools that integrate the conversion, flow, and value views'. In addition, they called for a collaborative and integrated exploration of the design management domain, in order to improve the conceptual framework, including the testing of potential tools, and the validation of hypotheses and theories. Building on Ballard and Koskela's (1998) previous work, Koskela *et al.* (2002) then published design management research involving empirical findings from their selected case studies. The findings suggested that there were still deficiencies in design management, in terms of process transformation (previously referred to as conversion), from inputs to outputs, the flow of information and the process of generating value for customers/ clients. Of interest was the Lean Project Delivery System as described in an updated article (Ballard, 2008), from the original publication (Ballard, 2000), which bears several complementary and close similarities of purpose/intent and process to design management: for example, working with the client to establish what they want to accomplish, and the constraints (location, cost and time) on the means of achieving their ends, and working with the other project players from the commencement of design work and therefore well before the project gets to the finalized documentation and on-site stages of the project. Ballard (2008) states that:

> In the Lean Project Delivery System, it is assumed that the job of the project delivery team is not only to provide what the customer wants, but to first help the customer decide what they want. Consequently, it is necessary to understand customer purpose and constraints, expose the customer to alternative means for accomplishing their purposes beyond those they have previously considered, and to help customers understand the consequences of their desires. This process inevitably changes all the variables: ends, means and constraints.

The following is a selection of key comments from the literature supportive of the development of a conceptual management model:

- 'Project management concepts and techniques have proven incapable of solving the difficult problem of design management. The main reason being that project management concepts are rooted in the conversion model' (Ballard and Koskela, 1998).
- 'There is a lack of a solid conceptual foundation (in terms of design management)' (Ballard and Koskela, 1998).
- There was 'a call for a collaborative and integrated exploration of the design management domain, in order to improve the conceptual framework, including the testing of potential tools, the validation of hypotheses and theories' (Koskela *et al.*, 2002).
- 'A call for new models and theories which recognise and illuminate the complexity of projects and their management at all levels' (Winter *et al.*, 2006).
- There was 'a lack of consideration of social skills in projects and the lack of contextualisation' (Geraldi *et al.*, 2008).
- There was 'a lack of effective information and knowledge dissemination and unsatisfactory management and performance levels and processes in current disaster management practices' (Haigh *et al.*, 2006).

REFLECTIONS ON THE METHODOLOGY/ PROCESS

Remote site design management was an unpublished phenomenon from a theoretical perspective when the doctoral research was conducted. Therefore, exploratory empirical research methods, including grounded theory, were considered to be useful developmental approaches. See Kestle (2009) for a fuller description of the methodology.

Grounded theory was adopted given its relevance to the research approach during the initial development stages of the theoretical conceptual model for remote site projects. Grounded theory was originally developed by Glaser and Strauss (1967) as a qualitative research method. Glaser (1978) described 'theoretical sensitivity' as being an important feature of grounded theory for researchers aiming to develop conceptual theory from first-hand expertise and experience in the research area. In addition, Strauss and Corbin (1994) suggested that meaningful and useful relationships between concepts result when using a grounded theory approach to conceptual theory development.

Historical research was also undertaken by Kestle (2009) at the exploratory stage and later at the model testing stages of the research process, using a document analysis approach. Historical research is identified as being useful when trying to explain, compare, predict and evaluate historical and current data and scenarios (Burns, 2000).

Finally, and again from the research by Kestle (2009), multi-case study methodology was adopted for the development and testing stages of the conceptual model and the analysis of the interview transcripts because it involved empirical enquiry that investigates a phenomenon within a real-life context. The phenomenon in this instance was remote site design management, which involved the study of remotely located collaborative international projects on environmentally sensitive world heritage sites.

The adopted multi-case study approach provides insights into processes, practical realities of theoretical premises, and whether the researcher's interpretations of issues, processes, data or theories are plausible (Burns, 2000; Creswell, 2002).

The objective at the exploratory research stage was to use a selection of primary and secondary data from earlier research, and reflections on personal experience around scientific base projects in the Ross Sea Region of Antarctica, Tongariro National Park New Zealand, and Kingfisher Bay Resort on Fraser Island, Australia. These data were then tabulated as benchmarking attributes and dimensions and analysed in terms of 'fitness for purpose' and relevance when developing the typology and conceptual model (Kestle *et al.*, 2002; Kestle, 2009). Following the development of the conceptual design management model, the research moved into the testing stage, to establish whether the theoretical model was rigorous and capable of representing a diverse range of current and completed remote site project contexts.

Two of the projects selected to test the model were the Cape Roberts Drilling Project in Antarctica and the UN Humanitarian Aid Project in West Darfur, Sudan (2004 onwards). The Cape Roberts Drilling Project case study involved a retrospective historical case study approach, reviewing previously published data (Cowie, 2002) and interviewing selected participants who had previously been involved on the project. The UN Humanitarian Aid Project case study in West Darfur, Sudan (2004) was selected as an active and current project, and a range of management and managed personnel were interviewed. The sampling approach involved selecting a cross-section of personnel who were representative of the various roles on both of the case studies. Semi-structured interviews were conducted with senior managers, middle managers and operational staff in terms of their official roles on the projects to establish the realities of in-the-field operations and management of these projects. The interviews were audio-taped and painstakingly transcribed, 'member checks' were conducted when requested by the participants, and replication logic was applied in order that the collected data and analysis of those data would be credible, valid and dependable (Kestle, 2009). An extensive analysis was then conducted involving each of the individual case study findings, and comparisons were made between the case studies' findings, and with the overall literature findings.

REFLECTIONS ON THE DEVELOPMENT OF THE CONCEPTUAL DESIGN MANAGEMENT MODEL AND THE LINKS WITH LEAN/DESIGN MANAGEMENT INTEGRATION

One of the significant outcomes of the review and analysis of lean design management and design management literature was the important contribution of process integration and value generation to the development of the conceptual design management model for remote sites.

CONTRIBUTIONS TO THE DEVELOPMENT OF THE MODEL FROM THE EXPLORATORY PROJECT CASE STUDIES

The projects selected to inform the early development of the conceptual design management model comprised Ross Sea Region scientific bases in Antarctica, an eco-tourism resort on Fraser Island, Australia, and Tongariro National Park huts and ski lodge sites, New Zealand. These projects/sites fitted the sociologically oriented holistic perspective to varying degrees while also drawing from the production-oriented worldview. This process aided in establishing the probable key factors or drivers to be considered in the development of the conceptual design management model for remote sites.

From the primary and secondary data and reviewed literature, at the exploratory stage of the research:

- the characteristics of remote sites were categorized, in the typology developed by Kestle *et al*. (2002), as being remote in relation to:
  - the distance of the site from continuously available logistical support;
  - the hostility of the environment in terms of the climate;
  - the difficulty of having physical access to the sites;
  - being located in areas of hostile physical conditions and environmental sensitivity;
  - a lack of available local materials and labour resources;

and

- the range of properties and attributes evidenced on the selected exploratory stage remote site projects included:
  - pristine environment;
  - governmental monitoring;
  - government as client;
  - evolving and mature ecosystems;
  - commercial value;
  - design stakeholders;
  - scientifically investigative activities;
  - global impact and historical conservation;
  - resource-rich;
  - hostile climate;
  - indigenous history.

The comparative discussion of the typological attribute findings suggested that there were in fact attributes in common between the selected remote sites that were not evident when they were initially selected. The data from the three exploratory case study remoter site projects provided the initial indicators of key criteria that form the basis of client and stakeholder priorities when developing projects on these remote sites. This, in conjunction with the other noted findings, while constructing the typology, informed the early development of the conceptual design management model for remote sites (Kestle, 2009).

For example, the added dimension of remote site locations tends to increase the complexity and critical timing of early decision-making. The resultant delays in making key decisions can mean that the entire project becomes unviable on remote sites, particularly where accessibility and constructability are limited by seasonal weather conditions and where scientific projects, for example, are funded to a strict 12-month timeline for the completion of a project. The timing of the decision to proceed towards the design and to financially commit to the project is absolutely critical to the detailed design and construction stages and completion of the project on time.

The three exploratory case study remote site projects were found to closely align with lean and functional production processes, as the main priorities for the client were shelter, a strict budget, tight timelines and a process-driven construction programme. The development of these projects/sites then potentially fits the 'traditional lean thinking design management model', in terms of the sequential process and flow approach. However, all development projects on Antarctic sites have to fully comply with the Environmental Protocol (1991), associated with the Antarctic Treaty (1959), and the Environmental Protection Act (1994), particularly in terms of minimizing environmental impacts. This meant in effect that the traditional lean design management approach did not fully address all of the factors associated with remote site design management (Kestle, 2009).

However, the design management practice and literature were also reviewed in terms of 'production principles' and 'sociological factors', and the 'characteristics/attributes and dimensions of remote sites' were referenced while developing the conceptual model (see Figure 1). The four factors and/or plausible drivers were arrived at by contextualizing the typological descriptors for remote sites, identifying the contributions made by

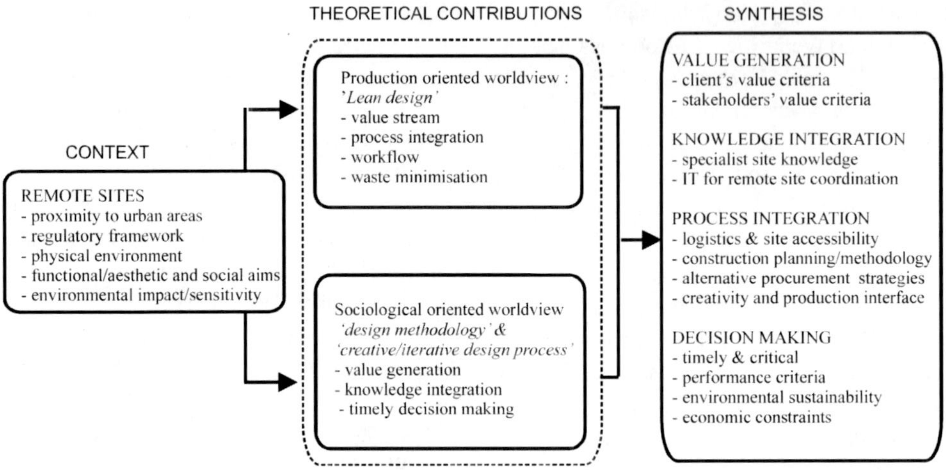

FIGURE 1 Exploratory design management conceptual model for remote sites (Kestle and London, 2002)

the sociological and production-oriented worldviews and producing a synthesis for the theoretical model (Kestle and London, 2002).

The principles and concepts of 'value generation', 'knowledge integration', 'timely decision-making' and 'process integration' became the key factors and potential drivers of the exploratory conceptual design management model for remote site projects.

SPECIFIC CONTEXTUALIZATIONS OF THE MODEL'S KEY FACTORS

Value generation refers to the value that the specific clients and stakeholders place on the project and the site, and they need to see this value (which is often intrinsic as well as economic) realized in the completed project. Value generation in this context is primarily concerned with the environmental protection of the site, where there is restricted public accessibility to the site, and with the site's global world heritage and/or esteem value.

Knowledge integration refers to specialist knowledge of remote (and often environmentally/climatically hostile) sites. Knowledge capital is a particularly valuable asset on remote site projects. The integration of this specialist knowledge is important to the decision-making and design choices at the pre-planning stages, and for the detailed pre-planning of all the stages or phases.

Timely decision-making in this context refers in the main to financial and design decisions, which are critical to the successful management of the design, construction and implementation of remote site projects. These decisions are frequently made within the context of non-negotiable windows of physical accessibility to the site(s), buildability, fixed budgetary constraints, political agendas and/or directives, and the need for environmentally sensitive development of these remotely located, pristine and often hostile sites.

Process integration involves several aspects, ranging from construction planning methodology, logistics, information management and the influence that the creative design stages can have on overall process management of the total project. Logistical planning and implementation is complex and critical for remote sites, often requiring up to a year of pre-planning, for example in Antarctica where access to sites is limited to a 4-month window and life-threatening situations are the norm.

When synthesizing the various theoretical and contextual contributions from the literature and practice, the exploratory conceptual design management model was created, which responded to, and reflected, the need for well-integrated specialist design, construction and operational factors (Kestle, 2009).

A further and relevant observation for this remote site management research was the comment made in the research network findings, by Winter *et al.* (2006), that 'theories about practice can also be used as theories for practice'. The network project also established from the data analysis that future research needed to specifically focus on creating theories and concepts that were closely aligned with 'in-the-field' realities and that provided project practitioners with realistic and contemporary management frameworks. The data collected from the practitioners by the research network referred to the complexity of projects, created in the main by 'the multiplicity of stakeholders and their differing agendas'. They suggested that concepts and frameworks that would help them deal with the project complexity issues 'in the midst of practice', and that recognized an interdisciplinary approach, social process, project conceptualization, value creation and value management, would be useful (Winter *et al.*, 2006).

TESTING THE RIGOUR AND OPERATIONAL USEFULNESS OF THE MODEL

USING THE ANTARCTIC CAPE ROBERTS DRILLING PROJECT CASE STUDY

This project was an international collaboration between seven nations in the Ross Sea Region of Antarctica. An article by Kestle and Storey (2005) reviewed a selection of the collected data from interviews conducted with nine participants who were drawn from a cross-section of the original management and managed personnel (including logistic and scientific). The aim was to establish how well the model represented the realities of managing the Cape Roberts Drilling Project in Antarctica and whether the gathered data supported, challenged or added to the four key factors of the conceptual design management model. The 'key factor' results were extensive and generally consistent across all of the nine selected participants. The following key points in Table 1 were drawn from those collected data:

One of the interesting outcomes from the collected data on the 'key factors of the model' was the range of views held by the interviewees regarding who they believed were the stakeholders and the client(s) on the Cape Roberts Project. Views were diverse. There was consensus, however, regarding what the value generation criteria were for the Cape Roberts Drilling Project. The primary purpose in their view was to drill and recover high-quality, specific sedimentary cores to create globally significant (climatic) scientific outcomes.

Another finding was the suggestion that knowledge integration can be inhibited, to some extent, on these scientific projects, by the issue of intellectual property, which may, and possibly does, work against the sharing of intellectual or knowledge capital nationally and internationally. There was also evidence of a lack of succession planning across all disciplines and areas of expertise (Kestle, 2009).

Reviewing the collected data regarding process integration, there was strong support for clear, effective, regular communications, and the centralized storage and management of data, on future projects. Communications were considered critical on remote sites. Miscommunications occurred at times between the various stakeholders on and off site, caused perhaps by different interpretations of issues or by decisions being made remotely from the site itself and from each other. Time delays in terms of decision-making, whether in terms of the design or financial commitment, cause a flow-on effect across all disciplines and tasks. In the case of remote site projects, this could mean a delay of up to 12 months until the site becomes accessible again.

The data supported the notion that strategic decisions made during the briefing, conceptual design stages and pre-planning stages in particular can significantly impact upon the design, fabrication, construction, logistics and the value generated in order to meet the expectations of the clients and stakeholders, and the final outcomes (Kestle, 2009).

The international significance of the collaborative scientific research, innovative technical expertise and 'knowledge capital' associated with the Cape Roberts Project has helped place New Zealand in a position of strength in terms of influencing the Antarctic Treaty partners, and future collaborative

TABLE 1 Key points from the Cape Roberts Drilling Project collected data with reference to the four key conceptual design management factors (sourced from Kestle, 2009, Table 4)

THE FOUR KEY FACTORS	KEY FINDINGS
Value generation as perceived or needing to be realized on the Cape Roberts Drilling Project was:	With respect to the technical and scientific aspects, and specifically the scientific outcomes from examining the cores
	The value added through the scientific outcomes of the project, and the contributions on the international stage, e.g. climate change, Kyoto Protocol
	The 1.7 km of core provided at a very reasonable cost. This outcome was considered 'great value for money' by the six nations involved with NZ and in the fact that USA also saw NZ technology as 'great value for money'
	In the need for low environmental impacts by the project and its personnel
Knowledge integration as perceived or needing to be realized on the Cape Roberts Drilling Project was:	Intellectual property issues preventing knowledge integration happening – seen as an inhibiting factor
	Successional planning being the need for understudies for the people with key intellectual capital
	'Risk' on these projects is significant in terms of the personnel selected for the project(s). If you get the wrong people, it can break up the project
	Problems that arise with 'patch protection', where people may not want to share their valuable expertise with potential successors for fear of becoming dispensable
	The situation where a pool of specialist remote site personnel is created within organizations to design and manage these particular project sites. The potential weakness, however, is where this knowledge capital is not documented explicitly and a successional framework of specialist personnel is not fully established within organizations
	Knowledge management using centralized databases and consequent data management makes the creation of a website essential on future projects
Process integration as perceived or needing to be realized on the Cape Roberts project and future Antarctic projects was:	The consequences of no process integration are dissatisfied staff, burn-out, budget blow-outs and an incomplete project
	This is all about operational logistics and information management, and it is critical that this occurs in a timely and realistic manner
	The secret is pre-planning and being aware of the other team members' needs and the consequences of all the actions proposed
Timely decision-making issues on the Cape Roberts and future Antarctic projects were that:	The key decision-makers have to be identified and recognized as having the appropriate authority to act and respond
	The fact that fast, accurate and safe decisions were made in potentially dangerous situations was the result of having a very good management structure that was responsive and responsible
	The participants noted that 'everything revolves around the environment on Antarctic project sites. The weather controls everything in terms of what, when, and if you can do anything. You have no real control; it's often called the Antarctic Factor'
	A lack of timely and critical decision-making may result in the loss of a whole year (or more) of core production, as ships can only access Antarctica, in the Ross Sea Region, once a year, between late September and February

international scientific endeavours. The fact that this was a collaborative venture meant that as with many collaborative working arrangements, there is always a tendency to make sure that the individual partner's needs and demands are met, as well as those associated with the agreed 'collective good'. This is where the situation can and did become political, on occasion. Working to tight timeframes on a project, providing internationally significant scientific data, set in a hostile climatic zone, with fragmented groups of project personnel, with differing expectations both financially and in terms of the value of the outcomes to them professionally, resulted in a complex, exciting (at times frustrating) and rewarding project.

The participants' data, once analysed, specifically supported the overall conceptual design management model, and particularly the four key factors, as being valid for Antarctic remote sites generally, and as accurately representing their experiences on the Cape Roberts Drilling Project (Kestle and Storey, 2005; Kestle, 2009).

USING THE UNITED NATIONS SUDANESE HUMANITARIAN AID PROJECT CASE STUDY

The United Nations Sudanese Humanitarian Aid (UNSHA) Project in West Darfur was an international collaboration, involving aid representatives from several countries (such as the UN HQs in Geneva and Khartoum, the USA, the UK, European Union, New Zealand and Australia). Each country contributed to the management and/or operational aspects of the project. This collaborative approach created a complex regime of project personnel and tasks that needed to be sensitively integrated, coordinated and managed.

An article by Kestle *et al.* (2006) focused specifically on testing the four key factors of the conceptual design management model for remote site projects on the UNSHA Project in West Darfur, using a selection of the collected data gathered from the eight middle and senior management interviewees. The factors were value generation, knowledge integration, process integration and timely decision-making.

The aim was to establish how well the gathered data supported, challenged or added to the four key factors of the conceptual design management model. The results (see Table 2) were generally consistent across all the selected interviewees, although some of the respondents appeared to have more autonomy than others in terms of playing a real part in the decision-making processes. The personnel interviewed supported the four key factors of the design management model as being valid for humanitarian aid project sites generally, and as being representative of their experiences, or as being those that were needed on projects such as the West Darfur Humanitarian Aid Project in Sudan.

The main challenge for all the aid agencies was establishing a means of quantifying whether the 'plan' had been achieved once implemented and whether the expected 'value' had been generated. A range of views emerged in terms of the preferred and actual process integration in practice on the project, and whether the respondents had to slavishly follow the plan from 'central' or whether local decision-making opportunities existed on the project. There was recurring criticism of the centralized decision-making process of some of the agencies, and how this hindered progress and timely communications and hindered the potential for on-the-ground, informed and improved/relevant local decisions from being made. Others believed that they had some autonomy in terms of the decision-making, having had the authority delegated to them by their agencies. This lack of consistency of decision-making and delegated authority, across the range of agencies, and the ever-changing personnel in-the-field and offices, was challenging, frustrating and disorientating for a number of the respondents. A commonly and strongly held view was that there were significant gaps in terms of specialist knowledge and knowledge integration on the West Darfur Humanitarian Aid Project, and insufficient pre-briefing and associated training, before going into the field. There was consensus among the respondents that there was a significant lack of effective communication equipment and systems available for project staff, in the field and in the offices at the start of the 'in-the-field' project work.

The eight selected interviewees were very clear that the conceptual model works, and unequivocally supported the four key factors as accurately representing their experiences, or those that were in fact needed on projects such as the UNSHA West

TABLE 2 Key points from the UNSHA West Darfur data with reference to the four key conceptual design management factors (sourced from Kestle, 2009, Table 5)

THE FOUR CONCEPTUAL DESIGN MANAGEMENT KEY FACTORS FOR REMOTE SITES	VALUE GENERATION AS PERCEIVED OR NEEDING TO BE REALIZED ON THE WEST DARFUR HUMANITARIAN AID PROJECT IN SUDAN AND AS PER THE COLLECTED DATA WAS:
Value generation	The effectiveness, and therefore the value, was measured on the project by what was achieved, how many internally displaced persons (IDPs) have been saved and fed and what the mortality rate was. Value was measured quantitatively
	About keeping a reliable, continuous supply line of food to the displaced people, from a distant donor to the NGOs in the field
	About making a difference to the living conditions, in terms of emergency water and sanitary assessments in the 'field', acting on the recommendations and their timely implementation
	Measured in how many built outputs will be achieved, and then seeing the re-collection of people; putting the 'village' back together again
Knowledge integration	That there are definite gaps in the knowledge integration process. No one wants to trespass on others' areas and therefore be perceived as a possible hindrance to finding the best solution(s)
	That there is a problem with the planning and the reality. Very specialized personnel who come in cannot do what they are best at, as they have to follow a particular plan, and therefore not necessarily seeing the desired/potential 'results on the ground'
	That there are consultants, who are not in the UN system, who need to be advised of the potential pitfalls, when involved on these types of projects
	That there are basically informal and formal systems of knowledge integration
	The gaps in specialist knowledge, in terms of the experiences of the people in the field, versus those in the office – they were not always in-line at times
	That sometimes there is too much specialized knowledge on a project, and what is needed is a more holistic approach
	A good knowledge of the IDPs' cultural and value systems is needed, before commencing the on-site work
	The high turnover rate of people in these roles, so things were not recorded as much as they could have been. Important though, to understand the context of the project
Process integration	To try and understand how the IDPs think, and will act/respond, and then to try and set up the best processes and systems
	In trying to achieve co-ordination at the camp level, and engage in meaningful and useful relationship building with the international and IDP communities. Knowing the other agencies' plans means better facilitation
	That little could have been achieved without the Sudanese people and their expertise. They had valuable connections and networks within the community
	About co-ordination of the various groups on this project and helping working groups focus on the task in hand
	To make sure that assessments are correct. That a thorough, logical and sensible solution to the assessment findings is made. Then prepare a plan to address the challenges within the timeframe and the budget

Continued

TABLE 2 Continued

THE FOUR CONCEPTUAL DESIGN MANAGEMENT KEY FACTORS FOR REMOTE SITES	VALUE GENERATION AS PERCEIVED OR NEEDING TO BE REALIZED ON THE WEST DARFUR HUMANITARIAN AID PROJECT IN SUDAN AND AS PER THE COLLECTED DATA WAS:
Timely decision-making	That decision-making on this project was quite reactive and prescriptive. The detailed and bigger picture decisions were fed from the 'field' back to central, where the tailoring occurred, and the decisions, and plans, were fine-tuned
	A tiered system of decision-making. Consultative decisions were made. The people with the on-the-ground or with the bigger-picture knowledge worked together to work out the best answers, and decide what was feasible
	That decision-making involved a group of managers, one manager for each of the programmes, and it was essentially de-centralized
	That at the organizational level, the decision-making was de-centralized. There were considerable levels of co-ordination between West Darfur, Khartoum and the agency's head office. The staff were given almost total autonomy in the 'field', and dedicated organizational finance personnel to work with

Darfur Project. The analysis of the semi-structured interviews suggested that the conceptual design management model for remote sites (Kestle and London, 2002) is relevant in a non-profit and/or humanitarian aid context. In addition, the model allows for a blending of traditional and modern management methods (Kestle *et al.*, 2006; Kestle, 2009).

This support by the participants lent significant support to the model and to the associated typology for remote sites, and was especially pertinent given that humanitarian aid projects are probably one of the worst situations that one could select to test, also being very different in many ways to the Antarctic science project (Kestle, 2009).

CONCLUSIONS

The aim was to reflect on how the integration of lean design management and design management thinking informed and influenced a conceptual design management model developed for internationally collaborative remote site projects, and the level of rigour achieved 'in the field' when tested.

This aim was met by reflecting on the development and testing of the conceptual management model, from a postdoctoral perspective. First, a selection of the literature reviewed in the doctoral research and associated publications was revisited, and recent and relevant pointillisms of published research work conducted by lean design and architectural management researchers were included and integrated into the literature reflections. Second, the methodology and process associated with the model's development and testing using grounded theory, and multi-case study methodology, were reidentified and reviewed in order to demonstrate the influence of the integration of lean thinking, lean design management and design management on those outcomes.

When writing up the original doctoral research in terms of the data analysis from the findings for the main case studies, all of the participants agreed that the four key factors of the management model represented their experiences on the particular project that they had been involved on, and had been interviewed about for the doctoral research. On reflection, this suggests that the synthesis of the sociologically oriented worldview and the production-oriented worldview used to develop the model was supported. By inference then, the integration of the lean design and design management literature and practice definitely influenced and informed the development of the model, and was clearly supported by the operational in-the-field findings from two case studies used to test the developed model.

REFERENCES

Antarctic Treaty, 1959, Government Printer New Zealand, Wellington, NZ.

Ballard, G., 2000, 'Managing workflow on design projects', *Proceedings of CIB W96 Architectural Management, Atlanta, USA, 19–20 May*.

Ballard, G., 2006, 'Rethinking project definition in terms of target costing', *Proceedings of the 14th Annual Conference of the International Group for Lean Construction (IGLC-14), Santiago, Chile, 25–27 July*, 77–89.

Ballard, G., 2008, 'The lean project delivery system – an update', *Lean Construction Journal* **1**(1), 1–19.

Ballard, G. and Koskela, L., 1998, 'On the agenda of design management research', *Proceedings of the 6th Annual Conference of the International Group for Lean Construction (IGLC-6), Guaruja, Sao Paulo, Brazil, 13–15 August*.

Burns, R.B., 2000, *Introduction to Research Methods*, Sage Publications, Thousand Oaks, CA.

Cowie, J. (ed), 2002, *Cape Roberts Project – Final Report 1995–2001*, The Antarctic Institute, Christchurch, NZ.

Creswell, J.W., 2002, *Educational Research – Planning, Conducting and Evaluating Quantitative and Qualitative Research*, Merrill Prentice Hall, New Jersey.

Emmitt, S., Sander, D. and Christoffersen, A.K., 2004, 'Implementing value through lean design management', *Proceedings of the 12th Annual Conference of the International Group for Lean Construction (IGLC-12), Elsinor, Denmark, 2–6 August*, 1–13.

Environmental Protection Act, 1994, Government Printer, Wellington, NZ.

Environmental Protocol, 1991, Government Printer, Wellington, NZ.

Formoso, C.T., Tzortzopoulos-Fazenda, P., Jobim, S.S. and Liedtke, R., 1998, 'Developing a protocol for managing the design process in the building industry', *Proceedings of the 6th Annual Conference of the International Group for Lean Construction (IGLC-6), Guaruja, Sao Paulo, Brazil, 13–15 August*.

Freire, J. and Alarcon, L.F., 2000, 'Achieving a lean design process', *Proceedings of the 8th Annual Conference of the International Group for Lean Construction (IGLC-8), Brighton, UK, 17–19 July*.

Garnett, N.A., 1999, 'Developing lean thinking in construction: a naturalistic lean construction', *Proceedings of the 7th Annual Conference of the International Group for Lean Constitution (IGLC-7), Berkeley, CA, USA, 26–28 July*.

Geraldi, J.G., Turner, J.R., Maylor, H., Soderholm, A., Hobday, M. and Brady, T., 2008, 'Innovation in project management: voices of researchers', *Journal of Project Management* **26**(5), 586–589.

Glaser, B., 1978, *Theoretical Sensitivity*, Sociology Press, Mill Valley, CA.

Glaser, B. and Strauss, A., 1967. *The Discovery of Grounded Theory*, Aldine, Chicago, USA.

Green, S.D., 1994, 'Beyond value engineering: SMART value management for building projects', *International Journal of Project Management* **12**(1), 49–56.

Haigh, R., Amaratunga, D. and Kerimanginaye, K., 2006, 'An exploration of the construction industry's role in disaster preparedness', *Proceedings of the COBRA Conference, London, UK, 7–8 September*.

Howell, G.A. and Ballard, G., 1998, 'Implementing lean construction: understanding and action', *Proceedings of the 6th Annual Conference of the International Group for Lean Construction (IGLC-6), Guaruja, Sao Paulo, Brazil, 13–15 August*.

Huovila, P. and Koskela, L., 1998, 'Contribution of the principles of lean construction to meet the challenges of sustainable development', *Proceedings of the 6th Annual Conference of the International Group for Lean Construction (IGLC-6), Guaruja, Sao Paulo, Brazil, 13–15 August*.

Jorgensen, B. and Emmitt, S., 2009, 'Investigating the integration of design and construction from a lean perspective', *Lean Construction Journal* **9**(2), 225–240.

Kestle, L., 2009, 'The development and validation of a conceptual design management model for international collaborative remote site projects', Unpublished PhD thesis, University of Canterbury, Christchurch, New Zealand.

Kestle, L. and London, K., 2002, 'Towards the development of a conceptual design management model for remote sites', in: C. Formoso and G. Ballard (eds), *Proceedings of the 10th Annual Conference on Lean Construction (IGLC-10), Gramado, Brazil, Federal University of Rio Grande Do Sul, Brazil, 6–8 August*, 309–322.

Kestle, L. and Storey, B., 2005, 'The complexities of an Antarctic drilling project, and the fit with a conceptual design management model for remote sites', in: A.H. Boussabaine, J. Lewis, R.J. Kirkham and G.E.M. Jared (eds), *Proceedings of the 1st International Conference on Built Environment Complexity, University of Liverpool, Liverpool, UK, 11–14 September*, 301–313.

Kestle, L., London, K., Bodger, P. and Storey, B., 2002, 'The development of a typology for remote construction projects', *Proceedings of the 18th ARCOM Annual Conference*, ARCOM & University of Reading, *Newcastle-on-Tyne, UK, 2–4 September*, 305–316.

Kestle, L., Potangaroa, R.T. and Storey, B., 2006, 'Sustainable post-disaster reconstruction projects in remote locations, and the fit with a conceptual design management model', in: D. Alexander (CD Rom ed), *Proceedings of the 3rd International i-Rec Conference on Post-Disaster Reconstruction: Meeting Stakeholders Interests, University of Florence, Florence, Italy, 17–19 May*, 47–58.

Koskela, L., 2004, 'Moving on – beyond lean thinking', *Lean Construction Journal* **1**(1), 24–37.

Koskela, L., Huovila, P. and Leinonen, J., 2002, 'Design management in building construction: from theory to practice', *Journal of Construction Research* **3**(1), 1–16.

London, K., 2002, 'Design management model for performance based briefing', *Proceedings CIB W96 Architectural Management Conference, Hong Kong, 3–6 May*.

Salvatierra-Garrido, J., Pasquire, C. and Thorpe, T., 2010, 'Critical review of the concept of value in lean construction theory', *Proceedings of the 18th*

Annual Conference on Lean Construction (IGLC-18), Haifa, Israel, 18–19 July, 33–41.

Strauss, A. and Corbin, J., 1994, *Basics of Qualitative Research: Grounded Theory Procedures and Techniques*, SAGE Publications, London, UK.

Tombesi, P., 1997. 'Travels from Flatlands', Unpublished PhD thesis, University of California, Los Angeles.

Tzortzopoulos-Fazenda, P. and Formoso, C.T., 1999, 'Considerations on application of lean construction principles to design management', *Proceedings of 7th International Group for Lean Construction (IGLC-7), Berkeley University, Berkeley, CA, USA, 26–27 July*.

Winter, M., Smith, C., Morris, P. and Cicmil, S., 2006, 'Directions for future research in project management: the main findings of a UK government-funded research network', *International Journal of Project Management* **24**(1), 638–649.

Womack, J.P. and Jones, D.T., 1996, *Lean Thinking: Banish Waste and Create Wealth in Your Corporation*, Simon and Schuster, New York.

Index

Page numbers in *Italics* represent tables.
Page numbers in **Bold** represent figures.

REPRINT WORK TICKET

PART TYPE

cover

9 7 8 1

Request details

Customer	TANDF
Request ID	4118
Created	2026-09-23 07:26:25
Status	new
Request source	DBS 1
Reason	Spoilt on Hunkeler

Job details

Part description	Limp
Job description	Personal Transferable Skills in Accounting Edu
ISBN	9781138818309
Part type	cover

Production requirements

Quantity to reprint	1
Pages	192
Spine size	10
Final size (H × W)	246 × 174
Plate section	
Inner stock / GSM	Amber Preprint 80 gsm / 80
Cover stock / GSM	Cover Paper / 240
Lamination	Gloss

INCOMPLETE BATCH

BOOK TITLE

Personal Transferable Skills in Account
Education

ISBN NUMBER

9781138818309

9 7 8 1 1 3 8 8 1 8 3 0 9